The Management of Innovation

The Management of Innovation

R. C. Parker
Ashridge Management College

1807 1982

175 YEARS OF PUBLISHING

JOHN WILEY AND SONS

Chichester · New York · Brisbane · Toronto · Singapore

Library of Congress Cataloging in Publication Data:

Parker, R.C. (R. Charles)
   The management of innovation.

   Includes index.
   1. Technological innovations—Great Britain—Case
studies. I. Title.
HD45.P334   338'.06   82-2737
ISBN 0 471 10421 3   AACR2

British Library Cataloguing in Publication Data:

Parker, R.C.
   The management of innovation.
   1. Technological innovations
   I. Title
   658.5'7   HD45
   ISBN 0 471 10421 3

Typeset by Photo-Graphics, Yarcombe, Honiton, Devon
Printed in the United States of America

To my wife—Elizabeth Croisette

Acknowledgements

This book has two aims: to present eight case histories concerned with innovation within the manufacturing industries and to suggest ways in which the management of innovation may be improved. The study was proposed in 1975 by an Innovation and Creativity Committee which met under the aegis of the Council of Engineering Institutions (CEI) and, following an interval for fund-raising, the project was started in April 1977 at Ashridge Management College. It was funded mainly by the Department of Industry and the Foundation for Management Education, but additional support was given by the then National Research Development Corporation, Turner & Newall Limited, the Dowty Group Limited, RTZ Services Limited, BOC International Limited, the Davy Corporation Limited, Rubery Owen Holdings Limited, and the CEI. The author is indebted to these bodies and companies. The project ceased in July 1980, by which time some sixty companies had been visited and longitudinal investigations had been carried out in six. Two case histories were lodged with the Case Clearing House at Cranfield.

In August 1978 the James Clayton Bequest Fund, administered by The Institution of Mechanical Engineers, granted a Senior Clayton Fellowship for the purpose of completing three partially finished case histories and for producing three more, making eight in all. These additional studies were considered necessary in order to obtain confirmation of hypotheses that were emerging from the early part of the work. The support from this fund was greatly appreciated for, without it, publication of this book would not have been possible.

When regular visits are made to companies it is inevitable that personal relationships develop, and this makes it difficult to describe a company from a detached viewpoint. This difficulty fortunately disappeared when three engineers expressed an interest in the work and each volunteered to prepare two case histories based upon the many written reports and documents. Sincere thanks are therefore due to Mr R.G. Fall, OBE, Professor A.D. Baxter, and Dr J.P. Marsden. The author is also greatly indebted to Dr G. Hayward for his continued interest and advice. Grateful acknowledgement is also due to Mr V. Nolan who kindly organized two synectics sessions.

The British Institute of Management Foundation gave permission for the reproduction of the Lists of Guidelines in Chapters 1 and 2.

The eight companies showed considerable interest in the project and requests for information were invariably granted. Particular mention should be made of the six executives who gave helpful comments on the draft submissions and who generously allowed the names of their companies to appear, namely: Mr P.J. Culpitt, Mr T.J. Lawson, Mr R. West, Mr G.B. Marson, Mr G.O. Luff, and Mr G.G. Woodhead.

I am likewise indebted to Mr Philip Sadler, Principal of Ashridge Management College, for his continual interest and for reading and commenting on the final manuscript, to Mr A.J. Johnson for Appendix III page 149, and to Helen Davies for compiling the index.

Thanks are due to my secretary, Miss Elizabeth Pitcher, for showing great skill and patience in dealing with numerous alterations and amendments to the draft manuscripts.

Contents

Foreword

Innovation is perhaps the most difficult of management tasks. This is partly because achieving innovation in an organization of any size involves energizing a large number of people with qualities normally associated with individual genius rather than corporate excellence. These are qualities such as originality, imagination, vision, determination, and entrepreneurial drive. It is also difficult because the organization through which the innovative individual or group must work has usually been designed not to innovate but to routinize. Those very aspects of an organization structure which make for effectiveness in stable conditions—such as clear allocation of authority and responsibility, functional specialization, and a hierarchical structure of decision-making—can act as powerful obstacles to new ideas and fresh approaches. At the same time, achieving innovation, at company level and nationally, is the most urgent task facing management today. As the newly industrializing countries gradually take over the manufacture of traditional and staple products, and as international competition generally intensifies, so the road to survival for our industries lies in the successful development and marketing of new products.

Following a lifelong career in industry, culminating in the position of Technical Director with Ferodo, Charles Parker has, since 1977, dedicated himself to two tasks—the study of the process of innovation in British industry and the active promotion and facilitation of innovation in many companies. In so doing he has, with typical clear-sightedness, gone straight to the core of the British industrial malaise by tackling the issue which is simultaneously the most important and the most intractable. His work shows clearly the vital role played by the quality and style of management, particularly at the highest levels, and illustrates how neglect of the development of professionalism in management has been associated with failure to respond to changed market conditions. Dr Parker's analysis also highlights the importance of attitudes and social climate in achieving real and lasting success in innovation. In these industries in which technology plays a leading role—and scientists and technologists number strongly among top management—there is a tendency to place too much reliance on techniques and systems and to give too little attention to the human factor. This is in sharp contrast to the approach of, for example, Japanese management.

I was very pleased when Ashridge Management College was selected as the academic base for Dr Parker's research, and I am sure that this book, which represents one of the most important outcomes of his project, will be extremely useful both to practitioners and to students of the management process. The responsibility for innovation, however, and the power to achieve it lie in the hands of top management in industry, and not with management researchers or institutions of management education. The valuable lessons of this study will only be of benefit to society if practising managers translate them into action. Given the essentially practical nature of Charles Parker's approach, based on his own industrial experience, I believe that his influence on management practice will be considerable, and that he will derive his own sense of achievement from examples of the practical application of his findings.

PHILIP SADLER
Principal
Ashridge Management College

Introduction

During the last decade Europe and the USA have rightly paid much public attention to two major aspects of their economy:

(1) The deteriorating performance of many of the traditional industries; and
(2) The lack of investment in new and up-to-date plant for the improvement of industrial performance,

and there has been wide agreement that insufficient attention has been paid to the most important single factor in both—the need for innovative design of new products and the considerations which inhibit this.

It is important to note, however, that there are a number of notable exceptions to these criticisms. Some countries perform better than others, and manufacturing companies can be found in all countries, both large and small, which are as efficient as any, and so lead their fields in world markets. They make a disproportionately large contribution to the wealth of their countries. The concern, then, is for those industries which may have had a good record but, because of the effects of the world recession, now need new competitive products in order to maintain, or regain, their position in world markets. How can this be best obtained?

In practically all areas of industry there is ample technology available on which new products can be based; equally it cannot be said that there is any shortage of entrepreneurial ideas on the application of available technology. If ideas are not put forward spontaneously there are a number of well-established techniques available for the generation and examination of ideas, and these have been shown to be effective in finding a good solution to a problem or the means of meeting a requirement which can be defined.

An idea, however, must be evaluated and accepted if it is to be translated into a new product or process, and it is here that difficulties can begin. To make headway, the idea must reach someone in the organization who can carry out a thorough evaluation and, if accepted, deploy a great many more resources for development. Thus management is involved from the first.

This involvement covers much more than picking an idea for development. The first and most important requirement is to ensure free and simple internal communications with no blocks or filters. There is a natural inertia to change in any organization which can prevent ideas being transmitted and an absence of appropriate incentives can make matters worse. Top management must make its need for innovation known to, and accepted by, all levels and produce a general climate which encourages and does not suppress ideas.

The purpose of this study was to determine factors which either facilitated or hindered innovation, and its objective was defined as 'to stimulate increased innovative developments within manufacturing industries by carrying out practical case studies, with the co-operation of selected companies, of the good and bad factors controlling innovation within the organization'. The project, though diagnostic in execution, had in the end to become prescriptive.

Chapter 1 contains a critical assessment of the manner in which eight companies were seen to carry out their programmes on new-product innovation. Chapter 2 proposes a model for innovation which enables a company to be graded into one of six categories according to the nature of the technology associated with product development. Forty-six guidelines for product innovation are listed and are divided into six groups, three of which concern a company's corporate responsibilities and three the operation of research, design, and development. Not all guidelines apply to every company, and for four of the six categories it proved possible to identify a smaller, relevant number. Chapter 11 gives a summary of the factors which were seen to either help or hinder the innovative process. It is hoped that Chapters 1, 2, and 11 will be useful to industrialists who may only have time to scan the case histories.

Chapters 3 to 10 comprise the case histories and are intended for students researching into the innovation process and management students studying for post-graduate degrees. Teaching notes for the eight case histories, and suggestions for handling the material are available from the author. They are intended to be a guide to the selection of the most suitable case for an intended teaching purpose.

CHAPTER 1

Guidelines for Product Innovation

1.1 Purpose of Study

The purpose of this study is to show how manufacturing companies might improve their performance by recognizing the responsibilities of management for product innovation. Few industries can grow, or even survive, for long unless they meet successfully the challenge of advancing technology with a succession of improved products and processes. It was apparent from the beginning that the most important factor was the corporate attitude of the board, and the leadership given by chief executives. This forward planning was too often based on the extrapolation of historic trends of financial statistics and made little or no mention of new products. Although policy on innovation should be a board responsibility the innovation process involves numerous activities ranging from the generation of original ideas, through production, to marketing, and hence all management share a responsibility.

Innovation is perhaps the most difficult of management tasks: it involves every company function and its time-scale is rarely less than five years and may exceed ten. Nevertheless failure continually to integrate all innovative activities will result in even the most brilliant ideas for a new product or process becoming sterile.

To achieve the aim of this work it was planned to examine the many discrete sectors of business activities and so diagnose the causes of the failure of management to innovate. Despite the rich diversity of purpose, nature, and structure of companies it proved possible to isolate the important factors and to express them as managerial precepts for decision and action. Not surprisingly, the picture was complex. A company can only plan its future in the context of the present, and this inevitably creates difficulty in the allocation of human and material resources between conflicting needs.

The number of precepts was restricted to those which were considered imperative rather than essential, yet they numbered forty-four, which was considered to be far too many to be put into action by executives and managers. For the study to be of practical value it was deemed essential to find means whereby a selection could be made of only those maxims that were thought appropriate to an individual company's needs.

1

1.2   Choice of Companies

The first consideration in choosing the companies was to ensure that, between them, they encompassed all stages of the innovation process. This was necessary because the four-year term of the study was likely to be less than the average period which elapses between an idea for a new product and its market-launch. The second requirement was, again, related to the restricted duration of the project, and called for companies to be manifestly interested in, and committed to, innovation. Time spent in persuading management to innovate, although no doubt a potential source of useful information, would not be sufficiently relevant to the main objective—how innovation was managed. Similarly, companies in which conflict was apparent between directors, managers, or between members of the board and management, were excluded for the reason that progress would be slow or even absent. Reassurance was sought on two additional matters: that the chief executive approved of and welcomed the study and that the company could provide adequate resources for initiating and progressing the proposed innovation.

The basis of selection was to cover as wide a canvas as possible, in the belief that observations of good and bad management practice would then have a general validity. The hope was, therefore, to be that the project should be hosted by companies which, keen to innovate and willing to co-operate, would represent a wide range of sizes, type of ownership, products, structure, organization, and manufacturing processes.

The project was described in a major financial newspaper and within a few weeks some eighty companies sought information. Preliminary visits were made to sixty, from which eight were selected on the basis of information gained by interviewing members of the board and senior managers for longitudinal studies. The smallest firm in the sample employed less than fifty, and the largest, which employed 2000, was an autonomous subsidiary of a large international group. The manufactured products represented both high and low technology, and while two companies had access to central group research and engineering laboratories, others had neither research nor development departments. The product range included consumer products, scientific instruments, and engineering products; some were custom-built and others mass-produced. Among the many different management activities represented were factoring, subcontracting, requisition, mergers, divestment, and venture groups.

1.3   Methodology

The first few days were spent in interviewing directors and senior management in order to learn something of their experience, skills, and attitudes. Most executives and managers were observed to have a deep commitment to their responsibilities and, when asked for a brief history of their company and a review of the current situation, gave replies that were so comprehensive and coherent that neither comments nor questions were necessary. The sessions were unstructured and were most rewarding when it proved possible to avoid any kind of interruption. Questions clearly interrupted the interviewees' train of thought, and what had been an interesting continuous flow of information was rarely resumed.

The number of staff interviewed in this first phase ranged from eight to twelve in each company. A large amount of information was recorded and, although both duplication and occasional contradictions were noted, this mattered little in view of the wealth of data.

To facilitate comparison of one company with another a framework was required and an *aide-memoire* was drawn up as shown in Figure 1.1. Attempts were made to complete one for each company from recorded data. However, although it helped to identify important features in the company's style of management and climate, many aspects proved to be irrelevant.

Figure 1.1 *Aide-memoire* for initial interviews

I. Background

  (a) Directors
1. Company history
2. Company purpose and objectives
3. Corporate planning — (e.g. term, turnover, profits, gap analysis)
4. Financial data
5. Nature of existing products and markets
6. Style of management
7. Management techniques (e.g. job descriptions, salary scales, job assessments, etc.)
8. Number employed, staff ratio, number of qualified staff
9. Divisional structure
10. Status and experience of product champion/business manager
11. Characteristic of business portfolio

  (b) Senior managers
1. Perceived style of management
2. Formality of organization
3. Liaison between divisions
4. Conflict resolution
5. Staff morale
6. Staff misfits
7. Management/labour relations

II. Attitudinal factors
1. Level of top management support
2. Is there an explicit policy for innovation?
3. Are the risks associated with innovation understood?
4. Is there an appreciation within the company of the changes which may result from radical innovation?
5. Is the need for innovation seen to be urgent?
6. Is the management progressive and participative?
7. What are the qualities of interpersonal and interdepartmental relationships?

III. Managerial — techniques

1. How is the need for a new product/process to be determined?
2. Are sales forecasting techniques or market surveys used?
3. Are formal methods used for idea-generation (specific strategies and tactics, synectics, morphological analysis, check lists, synergism, etc.?)
4. Do feasibility studies include economics, finance, and technical aspects?
5. Are formal methods used to select development (e.g. linear or geometrical programming)?
6. Are design aids used (CAD, VA, VE, Pablo, etc.)?
7. Is there budgetary control?
8. Are planning methods (e.g. precedent networks) and monitoring techniques used?
9. Is resource-levelling practised?
10. What methods are employed to meet peak demands?
11. Are there ground-rules for stopping projects?
12. Is there a concept of separate development stages with appropriate assignment of responsibility?
13. Does the organization consciously promote internal flow of formal and informal information?
14. Do detailed arrangements exist for introducing innovation to the works (e.g. tooling, union involvement) and sales (e.g. validation tests, after-sales service)?
15. Are there formal methods of selecting from possible innovations best matches with both existing technical resources and market opportunities?

IV. Managerial—organizational

1. Characteristics of company organization (e.g. hierarchical, organic, or functional)
2. Is the R & D/D & D structured to give functional, project, matrix, or mixed project teams?
3. The precise organizational and committee structure
4. How are contrasting situations handled (e.g. fire-fighting to large radical innovations)?
5. Are there gatekeepers?
6. Is there strong coupling with external sources of information?
7. Are there provisions for prototype manufacture?
8. Are there means for promoting a continuous, formal feedback from customers?
9. Are there works and sales-complaint systems?
10. Can customer-collaboration be obtained with early field trials?
11. Are facilities provided so that ideas can be tried without serious interruption to production schedules?
12. Does the decision-maker have executive authority and access to all relevant information?
13. Are designers provided with appropriate facilities (e.g. detailing draughtsmen)?

V. Policy

1. Are the proposed innovations general, market-specific, or customer-specific?

2. Are the innovations market or technically orientated?
3. Is the R & D defensive, offensive, or imitative?
4. Are the projects high-risk, low-risk, or a mixture?
5. Are the goals likely to be wide in concept or narrow?
6. When seeking new opportunities does the company tend to be inward-looking with little external contact or outward-looking and synergistic?
7. Do the innovations concern process, product, or both?
8. Balance of emphasis between established and new products
9. Balance between immediate, medium- and long-term growth
10. Is the strategy general, market-specific, or customer-specific?
11. Is consideration given to other than in-house innovations (e.g. licensing, acquisitions, joint ventures, technology transfer)?
12. Is there sympathy for intuitive ideas about market reactions?
13. Do the proposed innovations mainly lie within existing technical boundaries and are they sophisticated or low cost/unit weight products?
14. At what are the process innovations primarily aimed (e.g. lowering product costs, eliminating labour, improving reliability)?
15. Is there to be an apportionment of effort between fire-fighting and radical innovation?
16. Are the manpowered, skills, and financial resources adequate to support the innovation and its adoption by works and selling/service costs?
17. Is there emphasis on service, price, and quality?
18. Has the R & D/D & D been given high status?
19. Is there an awareness that vacillation may hand success to a customer?

A formal attitude survey was carried out in three companies, but the additional insights and knowledge gained were judged insufficient to justify the additional two days required and the exercise was not continued.

The next phase of the study comprised two different approaches. In four companies help was given to setting up an organization and climate in which new ideas could be generated and encouraged. In the other four the study was largely observational: development meetings were often attended and regular visits were made over a period of up to three years. With one exception the executives showed keen interest throughout the work and offered every possible facility and help.

Approximately halfway through the work it was decided to test the feasibility of writing guidelines based upon experience already gained within companies and upon the results of published researches into innovation. Two advantages were sought: a more compact framework for assessing a company's performance and the fulfilment of at least part of the study. It was found that sufficient data had been gathered to formulate eight sets of six or seven guidelines on the management of product innovation.[1] They were mainly derived from observations in firms and excluded external economic and fiscal matters.

Six of the eight sets of guidelines corresponded to discrete levels of management, and were presented in this way because a course of action is not readily accepted when its initiation embraces more than one level of management and spans different levels of responsibilities. Discussion and negotiations across interdepartmental and interdivisional barriers need a detached approach and careful preparation.

Towards the end of the study case histories were written for the eight companies and data collected on the preliminary visits to sixty companies was reviewed to give as complete a picture as possible.

1.4 Guideline and the Case Histories

The first set of six guidelines were titled 'Innovation and Group Head-quarters'; however, for the purpose of this analysis all but two of these are omitted because, although three subsidiaries of large groups were studied, little contact was made with the parent boards. The two retained guidelines are included in the second of the original set but now retitled 'Innovation and the Company Board'. The remaining seven of the original eight sets are rearranged to give one group of three concerned with corporate responsi-bilities and actions, and a second group of three dealing with the functional operation of research, design, and development. The number in each set ranges from seven to nine.

Because the guidelines were written halfway through the life of the project it was necessary to examine the completed case histories to see whether the original list needed additions, deletions, or amendments. It was also considered necessary to decide how many of the forty-seven guidelines were relevant to each company.

1.5 Guidelines and Corporate Behaviour

The first three sets of guidelines deal with the corporate responsibilities of a company, and Figures 1.2 to 1.4 indicate the extent to which individual guidelines were observed to be applicable to each of the eight companies. A guideline was said to be 'adopted' when the company was judged to have made a conscious decision to accept the management principle involved; 'adopted in part' indicates that a company's practices approximated to the intent of the guideline although not necessarily reflecting a formal decision; and 'not applicable' is used when a guideline is considered to be irrelevant to a particular company's business. Where an entry is left blank there was insufficient material in the case history on which to make a judgement.

A number of guidelines, e.g. Nos ii and iv, Figure 1.2, comprise two or even three separate parts. In these, however, the activities are so closely interrelated that it would not be sensible or, in some instances, logical to omit even one part.

Guideline	Company							
	1	2	3	4	5	6	7	8
i Express company purpose in a generic form thereby extending awareness of possible growth in related markets								N.A.
ii Review environmental factors and forecast market needs that are likely to prevail when new products are expected to be ready for sale								
iii Consider all company products and classify them into their business sector prospects and associated competitive attributes. It is convenient to construct a matrix								N.A.
iv Review the group's strengths and weaknesses, appraise likely competitors' activities, and issue brief corporate plan indicating likely growth areas								
v Consider the implication of the above examination and for each product decide whether it should be phased out of production, continue unchanged, or earmarked for development								N.A.
vi Decide on possible new market segments and export territories for growth, and list the actions needed to gain business from main competitors								
vii Review the various possibilities for new products, i.e. in-house, development, licensing arrangements, joint ventures, and acquisitions, and consider possible divestments								
viii Classify activities according to whether they are service projects or projects with a high or low probability of success. Draw up budgets for each class and monitor capital and revenue expenditure								N.A.
ix Allocate disposable company funds, plus non-trading income and loans between marketing, manufacturing, and product innovation, and formulate short-, medium- and long-term product strategies								

Key: Adopted Not adopted
 Adopted in part N.A. Not applicable

Figure 1.2 Guidelines. Innovation and the company board

Guideline	Company							
	1	2	3	4	5	6	7	8
i Draw up a list of design and development modifications for products whose market share needs increasing	N.A.	Adopted in part	Adopted	Adopted	Adopted in part	Adopted	Adopted in part	N.A.
ii Determine areas in which innovative changes to manufacturing processes and production engineering will lower product cost and improve performance	Adopted in part	Adopted in part	Adopted	Adopted	Adopted in part	Adopted	Adopted in part	N.A.
iii Investigate selling, distribution, and pricing policies, and seek ideas for a higher growth and larger market share	Adopted	Adopted	Adopted	Adopted	Adopted	Adopted	Not adopted	Adopted
iv Obtain innovative ideas for new products to meet unsatisfied market needs and ideas likely to create a need	Adopted	Not adopted	Adopted in part	Adopted	Adopted in part	Adopted	Not adopted	N.A.
v Group the ideas according to whether they concern selling and distribution, cost-saving, product-improvement, or a new product	Adopted	Not adopted	Adopted in part	Adopted	Adopted in part	Not adopted	Adopted in part	N.A.
vi Forecast for each idea the expected financial benefit, cost of achievement, and the time needed for completion	Adopted	Adopted in part	Adopted in part	Adopted	Adopted in part	Not adopted	Adopted	Adopted
vii Select the best idea, on the basis of available resources, and their benefit/cost ratio, and incorporate them in a board strategy that is in line with the group corporate plan	Not adopted	Not adopted	Not adopted	Not adopted	Not adopted	Not adopted	Not adopted	Not adopted

Key: Adopted   Not adopted
      Adopted in part   N.A. Not applicable

Figure 1.3   Guidelines. Innovation and company directors

Guideline	1	2	3	4	5	6	7	8
i Arrange for the innovation plan to be sponsored by a board member who will be seen to have influence and enthusiasm								N.A.
ii Tell employees of the plan so that they know the likely time-scale, probable risks, and the radical changes which may follow								N.A.
iii Maintain the fewest number of management levels, remembering that an innovating organization needs an organic rather than an hierarchical structure	N.A.							
iv See that the salary and status expectations of the non-line specialist are in no way inferior to those of line management	N.A.							
v Evaluate staff abilities and contrive to make use of their full potential, and provide opportunities for continual growth								
vi Create an awareness that inspiration is important and heighten the creative ability of staff by suitable training								
vii Set up synectics groups for the purpose of formulating new and feasible opportunities								N.A.
viii In designing the organization and planning the layout of offices and laboratories, heed must be paid to those factors which control the flow of necessary information								
ix Encourage an atmosphere of enthusiasm								

Key: Adopted   Not adopted

      Adopted in part   N.A. Not applicable

Figure 1.4   Guidelines. Innovation and company environment

It may be thought difficult to decide whether or not the more subjective guidelines can be adopted. Is it possible, for example, to decide the degree to which a company can appoint designers of the highest competence (No. ii, Figure 1.7)? The belief is that it can, because an acute awareness of attitudes is acquired when close personal relationships have developed from regularly meeting staff over a period of years.

Five out of the eight companies complied, either fully or partially, with no less than eighteen out of twenty-five guidelines. This picture may seem unexpectedly favourable until it is remembered that the case histories refer to companies who were selected because of their predisposition towards innovation. It is pertinent to note, at this point, that no less than seven companies were seen to encourage, wholly or in part, an atmosphere of enthusiasm, much of which was directed to product innovation.

Observations on the relevance of guidelines Nos i, iv, v, and vi in Figure 1.2 indicated wide use of corporate planning. This was also found to be the case for many of the fifty-two companies visited, though not selected, for case histories. This applied particularly to the smaller companies in which the owner, or one or more of the directors, had previous experience in large organizations. The procedures with which they were familiar were applied with considerable skill in that they were a simplified and flexible version appropriate to scaled-down activities.

A number of guidelines which similarly showed a high degree of acceptance are those that refer to actions aimed at increasing sales (Nos i, iii, and v, Figure 1.3) and is, no doubt, an inevitable consequence of a severe economic recession.

There was little indication that corporate planning, when it was practised, included a quantitative appraisal of actions needed for the development of new products. No conscious decisions were taken in the allocation of funds for future activities, nor was a formal assessment made of possible ideas for new projects, and when arbitrary decisions were made costs were not budgeted and progress not monitored. (See viii and ix, Figure 1.2; vii, Figure 1.3). In part-mitigation of this criticism it is fair to observe that companies Nos 4 and 6 decided that they had only one option and that was to achieve a major innovative improvement to their basic product.

An innovative environment is characterized by a wealth of new ideas and an eagerness to deal with, and benefit from, the process of change. An enthusiasm for change is helpful but it must be directed towards the recognition and reward of creative actions: it should echo Plato's dictum 'What is honoured in a country will be created there'. Creativity is too often regarded as a virtue appropriate to research laboratories rather than an activity which should permeate a company. It was therefore disappointing to discover that firms rarely consciously directed effort towards enhancing creative abilities (guidelines vi and vii, Figure 1.4). The two synectic sessions, described in case histories Nos 1 and 5, were instigated on the suggestion of the author, and this technique was not previously known to the companies.

Numerous methods for improving recognition of opportunities and problem-solving are now described in the literature and, as synectics is one of many procedures, guideline vii, Figure 1.4 can be deleted and regarded as part of the preceding guideline vi.

Considerable research has been conducted into the importance to the innovative process of good internal and external communications and, in particular, there is strong support for the hypothesis that information which passes along informal channels is the crucial component. From discussions within the eight companies it was evident that there was little or no awareness of the many fascinating investigations on this subject.

1.6 Guidelines and Research, Design, and Development

The second set of three guidelines, Figures 1.5 to 1.7, which concerns research, design, and development, were not seen to be widely applicable. Indeed, if the management of innovation can be evaluated by observing the number of guidelines which are seen to be in agreement with observed practices, then the upper echelons of general management have a more professional approach to their activities than do those concerned with technology.

Of the 176 entries 43 per cent are shown to be 'adopted' or 'adopted in part', and compares with 68 per cent for the 200 entries in Figures 1.1 to 1.4. A part of this is accounted for by the larger number of 'not applicable' comments on the second set of guidelines, but, even when this is allowed for, the difference is still considerable: 60 per cent compared with 73 per cent. For companies where corporate policy included strong support for product innovation it was surprising to find research and development staff who were so often unfamiliar with investigation into the research process that has been carried out during the last decade. Many behavioural patterns, other than the unconscious use of informal information channels, have been investigated and, although the results are normally only reported in R & D literature, much of it could be used with advantage in other divisions of a company.

The apathy shown towards the subject of creativity (noted in section 1.5) was paralleled by a reluctance to adopt published procedures for facilitating evolutionary and innovative advances (guidelines i to iv, Figure 1.6), and whereas three companies occasionally applied value analysis and value engineering, none used functional cost analysis, morphological analysis, or any of the many methodologies aimed at improving design procedures (guideline vii, Figure 1.7).

Quantitative appraisal of profits did not feature in corporate plans, and it was therefore only to be expected that there was a reluctance for research, design, and development staff to use formal planning networks and resource-levelling techniques (guideline vi, Figure 1.5). Only company No. 4 made effective use of planning networks.

Guidelines which were most often fully or partially adopted were, for the

Guideline	Company 1	2	3	4	5	6	7	8
i — Secure the commercial exploitation of the output from research, design, and development activities by forming a team to deal with each group of related products. These teams should be made accountable for the development of new products	Adopted	Not adopted	Adopted in part	Adopted	Adopted	Adopted	Not adopted	Adopted
ii — Set up a product/process department responsible for prototype production, and a plant design department to provide manufacturing plant capable of meeting product specifications and target costs	N.A.	Not adopted	Adopted in part	Adopted	N.A.	Adopted	Adopted in part	Adopted
iii — Form support teams of scientists and engineers needed to service the above groups together with an administrative department responsible for servicing information, cost, planning, output, and other needs	N.A.	Not adopted	N.A.	Adopted	Adopted	Adopted	Not adopted	Adopted
iv — Use appropriate methodologies for the more functional tasks as appropriate, e.g. functional cost analysis, value engineering, morphological analysis	N.A.	Adopted	Adopted in part	Adopted	Adopted in part	Adopted in part	Adopted	Adopted
v — Take action on chosen ideas by constructing planning networks for all developments and by using resource-levelling techniques	Not adopted	Adopted	Adopted	Adopted	Not adopted	Adopted in part	Not adopted	Adopted
vi — Establish *ad hoc* working parties with the purpose of facilitating effective communication between company divisions, especially marketing, sales, manufacture, research, design, and development	N.A.	Not adopted	Adopted	Adopted	Adopted	Adopted	Not adopted	Adopted
vii — Where remote central research services exist, set up joint project teams and appoint a leader from the subsidiary company	N.A.	N.A.	N.A.	N.A.	N.A.	Adopted	Adopted	N.A.
viii — If use of the products involves expertise with which the customer is unfamiliar, provide training and after-sales service	N.A.	Adopted	Adopted	Adopted	Adopted	Adopted	Adopted	Adopted

Key:
Adopted   Not adopted
Adopted in part   N.A. Not applicable

Figure 1.5 Guidelines. Organizing research, design, and development for innovation

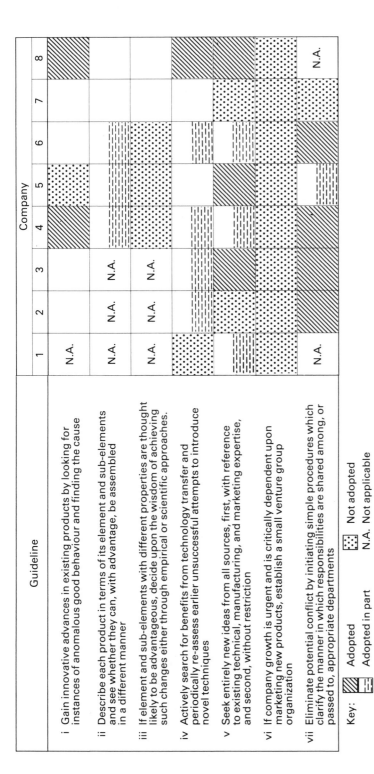

Figure 1.6 Guidelines. Innovation in research and development

Guideline	Company							
	1	2	3	4	5	6	7	8
i The head of design function must be a senior executive who co-ordinates all activities relevant to the product, especially the marketing, production, and finance functions	N.A.		N.A.					
ii Appoint designers of the highest possible competence	N.A.							
iii Expertise will be required in the following facets of engineering: industrial, materials, maintenance, reliability, safety, quality assurance, and purchasing	N.A.							
iv Cultivate an awareness of current technologies and methodologies, and take cognisance of psychological researches into creative thinking								N.A.
v A retrieval system must be designed whereby every previous design part can be located and adapted wherever possible	N.A.							
vi An engineering design and a design detail office should be set up, the first dealing with concepts conveyed through drawings or sketches, and the second meeting the needs of the various activities listed in (iii) above	N.A.							
vii For large projects, institute a separate design scrutiny function to guard against mistakes that arise through overfamiliarity with a design	N.A.							

Key: Adopted   Not adopted
      Adopted in part   N.A. Not applicable

Figure 1.7 Guidelines. Innovation and design

most part, concerned with organization, e.g. the setting up of departments and teams (guidelines i and ii, Figure 1.5) to assist the formal flow of information between company divisions. The interest in organizational matters was confirmed by the fact that no less than eight companies provided training and after-sales service (guideline viii, Figure 1.5).

One half of the entries in Figure 1.7 are either 'not applicable' or are left blank. This is partly explained by a failure of companies to recognize that design should be a separate activity from research and development. It is too often done badly, and so leads to difficulties with prototype-building, failure to meet customers' applications, and shortcomings with quality, reliability, durability, and maintenance. Whenever possible, design should be the direct responsibility of a senior executive. One cause of this neglect arises from the fact that the major output of many firms is not an end-product but component parts which have to be manufactured to tight specifications which leave little freedom for design variations. If the firm's remaining products are then regarded as a less important activity there will be a reluctance to appoint design staff of a high standard.

1.7   General Observations

The group of which company No. 5 was part had a central research laboratory, and the large international group to which company No. 7 was attached had a central engineering laboratory. Company No. 5 used its central facility to good purpose, but company No. 7 did not. The extent to which two organizations maintain good communications has been shown to lessen with time, and to be a function of the distance between them. Company No. 5 overcame the danger of collaboration decaying with time by setting up a joint product team, and were fortunate in that the factory and central laboratory were only a short distance from each other. Company No. 7, on the other hand, were separated from the engineering laboratory by a cross-country journey of several hours' duration, and liaison never reached the stage of sharing a project. The method of financing the central facilities also favoured the first of the two companies. Each subsidiary's contribution consisted of a fixed element, which could be used either on a short- or long-term project according to perceived needs, and a variable element which was a payment for contract work. The central engineering laboratory was less fortunate in that it was required to operate as a profit centre on an equal basis with the manufacturing units. It was therefore unable to compete with many of the subsidized technical services offered by a number of government departments and technical institutions. It could, therefore, only maintain its operations through general subcontract work, much of which was foreign to the needs of the group.

One guideline in the original set read 'Ensure that chief executives in each company exercise leadership and make them accountable for the achievement of growth targets'. It was not included in these guidelines because the study did not include visits to group headquarters. It is, nevertheless, clear from the case histories that leadership is the most important of all factors. Leadership was easy to recognize from comments to the effect 'We could not have done this before the present managing director was appointed'—a comment which was made, in various guises, by all levels of management. Good leadership was seen to produce an expectedly high performance by employees and, moreover, was a strong motivating influence and evoked an atmosphere of enthusiasm. An observation on leadership by Clausewitz (see reference 2) seems apposite: 'The more a general demands of his troops the better his demands will be met.' Among many attributes of leadership the study indicates the importance of integrity, sincerity, commitment, capacity for hard work, and approachability. It is perhaps worth noting that the two subsidiaries operating under high-quality professional leadership were alone in paying attention to staff development and management succession.

Although it is tempting to use the entries in Figures 1.2 to 1.7 to rate the innovative abilities of the eight companies, to do so would be invidious and misleading; there are too few characteristics in common to justify comparisons being made. This is illustrated by the varying number of 'not applicable' entries which, for companies Nos 1, 8, and 6, totalled nineteen, fourteen, and one, respectively. Nor does it follow that the first two companies are more alike than, say, companies 1 and 6. Indeed, the converse is true, and the contrast between companies 1 and 8 is greater than for any other pair. The former supplies large quantities of a consumer product whose design and manufacture involves a comparatively low level of technology, whereas the latter makes few products of considerable sophistication. Company No. 1 only set up its development department towards the end of the study and the guidelines in Figures 1.5 to 1.7, therefore, were irrelevant. The products manufactured by company No. 8 dominated world markets and, because sales repeatedly exceeded even short-term forecasts, the board were hesitant to spend time on long-term corporate plans. Figures 1.2 to 1.5 were therefore regarded as not applicable, at least during the period of buoyant sales.

Ideally, every guideline should carry a message worthy of immediate attention but, even from as few as eight case histories, it is apparent that differing company situations call for different sets of guidelines. Indeed, were this not so, their usefulness would be suspect. Ways must therefore be found whereby the characteristics of a company may be described in a manner that facilitates a selection of only relevant guidelines. This is a matter which is dealt with in Chapter 2.

References

1. Parker, R.C. (1980). *Guidelines for Product Innovation*, London, British Institute of Management Foundation.
2. Widmer, H. (1980). 'Business Lessons from Military Strategy', *The McKinsey Quarterly*, pp. 59–67.

Bibliography

Fig. 1.2 (i) Drucker, P.F. (1974). *Management: Tasks, Responsibilities and Practices*, London, Heinemann.

Fig. 1.2 (ii) Catling, H. (1972). 'Conditions for Innovation— with Particular Reference to Textiles', *R & D Management*, **2**, No. 2, 75–82.

Fig. 1.2 (iii) Robinson, S.J.Q., Hichens, R.E., and Wade, D.P. (June 1978). 'The Directorial Policy Matrix—Tool for Strategic Planning', *Long Range Planning*, **11**, No. 3, 8–15.

Fig. 1.2 (iv) Allen, M.G. (Autumn 1978). 'Strategic Planning with a Competitive Focus', *McKinsey Quarterly*, 2–13.

Fig. 1.2 (v) Hedley, B. (February 1977). 'Strategy and the Business Portfolio', *Long Range Planning*, **10**, No. 1, 9–15.

Fig. 1.2 (vi) Mason, R.S. (1977). 'Product Maturity and Market Strategy', *Cranfield Management Review*, **2**, No. 1, 44–5.

Carson, J.W. and Rickards, T. (1979). 'Three-dimensional Representation of Company Business and Investigational Activities', *Industrial New Product Development*, **5**, No. 1, 35–40.

Fig. 1.2 (vii) Killing, J.P. (1978). 'Diversification through Licensing', *R & D Management*, **8**, No. 3, 159–65.

Fig. 1.2 (vii) Pearson, B. (May 1978). 'Acquisitions without Tears', *Accounts Weekly*, 22–4.

Fig. 1.2 (viii) Parker, R.C. (1971). 'Controlling R & D Projects by Networks', *R & D Management*, **1**, No. 3, 147–53.

Fig. 1.2 (ix) Cox, J.G. (1980). *Growth, Innovation and Employment: an Anglo–German Comparison*, London, Anglo–German Foundation for the Study of Industrial Society.

Fig. 1.3 (i) Schliksupp, H. (1977). 'Idea Generation for Industrial Firms', *R & D Management*, **7**, 61–9.

Fig. 1.3 (ii) Bowe, C. (1977). 'Improving Manufacturing Efficiency—a Case Study for Private Industry', *Industrial Efficiency and the Role of Government*, London, HMSO, pp. 207–21.

Fig. 1.3 (iv) Ward, P.E. (1980). 'Focusing Innovative Effort through a Convergent Dialogue', *Long Range Planning*. **13**, 32–41.

Fig. 1.3 (v) Parker, R.C. (1974). 'R & D Evaluation and Selection', *Institution of Mechanical Engineers, Conference Publication 12*, London, pp. 12–19.

Fig. 1.3 (vi) McRae, T.W. (March, April, and May, 1979). 'Finance, Product Innovation', *Engineering*, 323–5 (March); 440–7 (April); and 601–3 (May).

Fig. 1.4 (i) Rothwell, R. (1975). 'From Invention to New Business via the New Venture Approach', *Management Decision*, **13**, No. 1, 10–21.

Fig. 1.4 (ii) Crawford, C.M. (1977). 'Product Development: Today's Most Common Mistakes', *Business Review*, **29**, No. 1, 1–6.

Fig. 1.4 (iii) Ovland, P.L. (1974). 'Discussion on Principles of Organising Applied Research and Development', *Research Policy*, **2**, No. 4, 322–34.

Fig. 1.4 (vi) Parker, R.C. (1976). *The Nurturing of a Creative Atmosphere in an R & D Laboratory*, London, Institution of Mechanical Engineers, pp. 1–22.

Fig. 1.4 (vii) Parker, R.C. (February 1975). 'Creativity—a Case History', *Engineering*, pp. 1–5.

Fig. 1.4 (viii) Allen, T.J. (1978). *Managing the Flow of Technology*, Cambridge (Mass.), The MIT Press.

Fig. 1.5 (i) (iii) Parker, R.C. (1977). 'Human Aspects of R & D Organisation', *R & D Management*, **7**, No. 3, 167–72.

Fig. 1.5 (iv) Parker, R.C. (1971). 'Controlling R & D Projects by Networks', *R & D Management*, **1**, No. 3, 147–53.

Fig. 1.5 (vi) Crawford, C.M. (1977). 'Product Development: Today's most Common Mistakes', *Business Review,* **29**, No. 1, 1–6.

Fig. 1.5 (vii) Hough, E.A. (1972). 'Communication of Technical Information between Overseas Markets and Head Office Laboratories', *R & D Management,* **3**, No. 1, 1–5.

Fig. 1.6 (i) (ii) (iii) Parker, R.C. (1970–1). 'The Art and Science of Selecting and Solving Research and Development Problems', *Proceedings, Institution of Mechanical Engineers,* **185**, No. 64, 879–93.

Fig. 1.6 (iv) Rothwell, R., and Robertson, A.B. (October 1973). 'The Role of Communication in Technological Innovation, *Research Policy,* **2**, No. 3, 204–55.

Fig. 1.6 (vi) Roberts, E.B. (1980). 'New Ventures for Corporate Growth', *Harvard Business Review,* 134–41.

Fig. 1.6 (vi) Fast, N.D. (1979). 'Key Managerial Factors in New Venture Developments', *Industrial Marketing Management,* **8**, 221–35.

Fig. 1.7 (i) Pugh, S. (1977). *The Engineering Designer—his Task and Information Needs,* Southampton, University of Southampton, pp. 63–6.

Fig. 1.7 (iii) Conway, H. (1977). 'Design and Produce', *The Production Engineer,* **56**, 21–6.

Fig. 1.7 (iv) Turner, B.T. (1973). *Creativity in Engineering—an Overview of some Methodologies of Engineering Design Work,* London, Institution of Mechanical Engineers.

Fig. 1.7 (v) Hamilton, W.A., and Lloyd, M.W. (1976). *The Contribution of Engineering Design Standards to Design Products,* London, Institution of Mechanical Engineers.

Fig. 1.7 (vi) Frith, D. (1965). *On Being a Designer,* London, Industrial and Trade Fairs Ltd.

Fig. 1.7 (vii) Turner, B.T. (1977). *The Best Format for Design Information,* Southampton, University of Southampton, pp. 87–94.

Fig. 1.7 (vii) Turner, B.T., and Leech, D.J. (1981). 'Management of Engineering Change', *Chartered Mechanical Engineer,* 38–42.

CHAPTER 2

The Level of Technology—A Basic Determinant

2.1 The Commitment of a Company to Innovate

In choosing companies for the longitudinal studies a judgement was made on their commitment to innovate. The aim of the interviews was to assess top management's attitudes to change and their willingness to allocate or, if necessary, to acquire the human and material resources. However, not all the eight companies lived up to these expectations with regard to their intentions towards innovation. This was particularly true of those companies selected during the initial stages of the programme and suggested, not surprisingly, that the earlier choices were based on too few factors.

The reason why such judgements became more accurate and meaningful with time was not immediately apparent, but it suggested that, based on an unconscious process of collating the many observations, more reliable assessments were being made. This prompted the idea that if the nature of the critical factors could be recognized and, perhaps, precisely formulated, therein would lie the basis of a quicker and possibly more satisfactory method of assessing the potential for innovation of a manufacturing company. Once this could be done there would follow the likelihood that knowledge could be obtained about the more important relevant management factors.

In selecting the companies an additional decision had been taken to include a wide range of dissimilar manufactured products. This was because earlier observation had indicated that a measure of a company's ability to market new products depended on combinations of different factors. The aim, therefore, was to gather data over a wide field in order to see whether patterns could be arranged to give a model which would delineate the innovative capacity of a company and so indicate the relevant set of guidelines.

2.2 The Importance of Innovation

The perceived need to innovate varies between companies. Successful manufacturers of microelectronic components, computers, and biotechnology products are highly innovative and produce a steady flow of new products,

FUNCTION	POSITIONAL COMPANY	INNOVATIVE COMPANY
Board	Emphasis on financial control	Innovation-orientated — future perceived as uncertain
Organization	Impersonal, hierarchical, status-dependent	Dual structure — vertical and horizontal
Marketing	Reactive — stability based on attractiveness of product. Closed marketing strategy	Constructively creates an unstable environment
R & D	Defensive, evolutionary	Aggressive, innovative
Production	Efficiency, rationalization, and long runs	Openness to change

Figure 2.1   Some characteristics of positional and innovative companies

whereas firms manufacturing drinks and foodstuffs for the consumer market frequently enjoy long periods of maintained growth with few, if any significant additions to their ranges of products.

These contrasting types of operation are reflected in the ways they allocate their disposable company funds, non-trading income, and loans to the development of new products. For example, whereas companies in the new electronic industries may typically allocate 15 per cent to this purpose, the figure for manufacturers of drinks and foodstuffs rarely reaches 3 per cent. The suppliers of consumer goods to the mass market, however, spend considerably more on achieving a high level of production efficiency and assign high priority to distribution, sales, marketing, and advertising. The contrasting emphasis between these two classes of business interests reflects differences ranging over the whole field of business activity, and an examination of the more important differences will be made to see whether they can be used to construct a basis on which to build a model of a company with reference to its innovative capacity. Nyström[1] has described a basic classification model of innovation and non-innovation (which he terms positional) companies, and a selection of his findings are summarized in Figure 2.1.

A positional company necessarily operates in a stable environment, eschews change, and adopts a reactive attitude to market requirements. The manufacturing operations associated with relatively unchanging products are concerned with attaining the highest possible efficiency, and this they achieve by giving attention to automation, group technology, rationalization, and other production engineering procedures. If research and development or design is carried out it is, normally, evolutionary and defensive in character. The company organization, not unnaturally, tends to be impersonal, hierarchical, and status-dependent. The boards of positional companies can afford to lay stress on financial control and corporate plans may, with some justification, be based upon the use of extrapolated historic trends of financial statistics.

The innovative company accepts that the future is likely to be uncertain and the process of corporate planning must itself generate new insights and be dynamic and reiterative. One aim of the marketing function in an innovative company will be to disturb a stable environment in order to create new opportunities. Manufacturing operations will thus need to be designed with a view to the possibility of introducing new production methods. These will be more important than efficient production. Research, design, and development will be important activities and will be aggressive and innovative rather than reactive. The organization pattern is likely to have the character of a dual structure, a vertical one based on a specialized function and a horizontal one concerned with co-ordinating independent activities.

Of the eight companies whose case histories are described below, one may be seen to correspond to Nyström's innovative model (case history No. 8, Chapter 10) and one (case history No. 2, Chapter 4) proved to differ little from the positional classification. The remaining six can neither be classified as positional nor innovative and so point to the desirability of deriving a model which could fill in the middle ground between Nyström's two classifications.

The importance of innovation for business firms is now seldom disputed but little has been written about the nature and degree of innovation. One trend that has been noticeable during the recession of the late 1970s and early 1980s is that many companies of international repute have decided to adopt innovative policies after decades of a positional stance. The traditional strategy of achieving growth by marketing established products more effectively in worldwide markets too often attracted vigorous competition, which inevitably led to losses in turnover and profits. In building a model it will be necessary to seek general applicability so that suitable strategies can be formulated for companies who need to graft innovative policies onto the positional form, or otherwise alter their particular mixture of innovative and traditional products.

2.3 The Innovative and Evolutionary Behaviour of a Company is a Function of its Technology

The one highly innovative company (case history No. 8) exhibits many characteristics that are not shared by any of the other seven. The most notable difference lies in their use of advanced research which is carried out by scientists and technologists. No less than one in four of the employees are graduates and one in ten postgraduates. This reflects the important purpose of the company, which is to achieve international excellence and renown while working at the frontiers of technology. Research is carried out in several disciplines and, although the company's success and reputation was founded on its achievements in cryogenics and superconducting magnets, it has achieved more recent growth by gaining technical leadership in medical instruments and electronic components for other specialized applications. Partly because of the company's early history, in-house research has been continually nurtured by a policy of maintaining the closest possible contacts with developments carried out by the users of products and has, additionally, maintained links with relevant centres of scientific excellence.

A second distinguishing factor of the highly innovative firm stems from the fact that a high proportion of its turnover is attributable to sales of relatively few high-cost products. Innovative developments are aimed at achieving high rewards, and this entails considerable research cost and capital expenditure. The risk incurred in high technology is usually considered to be great, but the continued uninterrupted growth of this company demonstrates that one counter to uncertainty lies in the recruitment and retention of highly qualified staff.

As the case history shows, this company has had difficulties. At one period during its early growth severe cash-flow problems were experienced which were overcome only by the appointment of a physicist with a proven record in general management as managing director. He was able to marry commercial needs with an understanding of the scientific ethos.

Turning now to the relatively non-innovative company (case history No. 2), an immediate contrast is shown. Despite its very much larger number of employees, these include neither qualified scientists nor chartered engineers. The company was founded on one basic product, the design of which depends mainly on craftsmen's skills, experience, and ability to react quickly to the needs of the customer. More recent strategies have been aimed at designing and marketing engineering plant associated with the use of their products, and newly appointed technical staff are more highly qualified than hitherto. A second characteristic which is unique among this set of case histories is the nature of the manufacturing process, which produces low-cost articles at a rate of thousands of millions a year.

The company's future does not seem to depend on product innovation based on advanced technology but rather on its ability to maintain competitive prices, a willingness to evolve product-modification to meet changing needs, and attention to non-price factors such as service, distribution, and selling. In further contrast to the highly innovative company, product-evolution is inexpensive but potential rewards are modest. The need to be competitive in this and other non-innovative companies places emphasis on efficient manufacturing. The main threat to the company lies in greater efficiency being achieved by competitors and in special factors associated with imports from the Far East.

It should be noticed that the above differentiating features are, to a minor extent, beginning to be blurred by recent developments. The first company is moving a little down-market in one or two of its electronic components for the industrial market, and the second is trying to introduce new products based on higher technical skills and is establishing contact with universities and polytechnics. Nevertheless eight contrasting features remain: the scale of revenue and capital expenditure on research; the risk associated with new-product development; the anticipated benefits from product-development; production quantities and cost of products; the source of development ideas; the importance placed on research, design, and development; paths to growth; and likely threats to company growth.

Evolutionary development describes work aimed at continuous improvement of products or processes to meet slowly changing needs, or an uninterrupted assimilation of evolving science and technology directed towards sustaining or expanding existing markets. Innovation involves the birth of a new idea, often an invention, together with its successful progression towards a new material, process, product, or system. Above all, it implies a discontinuity and a need for a radical shift in the way in which a company should be managed. It is for this latter reason that the widespread use of the term 'incremental' innovation to replace evolution is deprecated. Evolutionary development is a feature of a positional company, and to replace it with a terminology that connotes a degree of radical change is both misleading and confusing. In justification of the use of the term 'incremental' it is often argued that the totality of a series of evolutionary modifications is likely to constitute a noteworthy innovation. This is true, but it will take place over a long period of time and so enable a company to make the necessary adjustments to its structure without undue disturbance.

For the nature and degree of activities involved in product-development to be used as a means of discriminating between companies' capacity for innovation, a numerical rating must be devised which will distinguish between as many stages as possible, from the positional to the innovative modes.

Three separate approaches were examined. The first concerned the nature of the problem-solving activity required to convert an idea into a new product. In case history No. 2, for example, the problems were largely solved by repetitive applications of past experience. For a long-established company a very considerable amount of practical knowledge is gained in this way, and may constitute a powerful deterrent to would-be competitors. The innovative company, on the other hand, has to deal with abstract concepts at the frontiers of knowledge and has the analytical ability to solve complex quantitative relationships. From an examination of the remaining four case histories it proved possible to distinguish between six levels of problem-solving activity.

The second approach was based on the observed level of skill and qualification of staff employed on new-product development by the eight companies. They range from craftsmen and technicians in company No. 2 to engineers and scientists of international status in company No. 6, who were capable of achieving precisely formulated, unambiguous, high-technology goals. Again it proved possible to define six steps.

The third approach was to adapt a five-point scale that Langrish, Gibbons,

Level of technology	Nature of problem-solving task required to convert an idea into of a new product	Staff required for new product development	Completion of the problem-solving tasks would justify	New product involves a process of
1	Repetitive solution from simple choice of things learnt	Craftsman	Only rare mention in publications	
2	Patterned. Discriminating choice from past experience and existing knowledge	Craftsman and technical	Mention in trade journals	Evolution
3	New ideas. Moderate level of uncertainty. Improvement main aim	Qualified engineer/scientist	Mention in technical journals	Evolution with some innovation
4	New products alien to production and marketing enterprise. Open-end problems with infinite number of possible solutions. High uncertainty	Highly experienced engineer/scientist	Publication of papers in scientific or technical journals	Some evolution with innovation
5	Adaptive. Discriminating choice of spin-off from high/medium technology	Engineer/scientist. National reputation	Publication of papers in 'prestige' journals and cause substantial modifications to textbook	
6	Precisely formulated, unambiguous high technological goals. New knowledge, power of abstract thinking. Often quantitative problems, and singular solution	Engineer/scientist. International reputation	Sufficient papers in 'prestige' journals to justify a new textbook	Innovation

Evans, and Jevons[2] devised to measure changes in technology. Their scale ranged from a change that made no (or only a slight) difference to a standard textbook, through alterations or additions to a few paragraphs in a new book, to a new title. The suggestion here is to employ a similar concept but to apply it to the descriptions of manufactured products and to use it as a measure of the level of technology. Again, from an examination of the case histories, six levels were distinguished. For example, the development of products described in company No. 3 would just merit a description in a technical article, whereas the scientific research referred to in Chapter 10 would merit a new textbook.

The proposed subdivisions of the problem-solving activity, the skills and qualifications of the staff, and the importance of an innovative development judged by whether and where it merited publication were not only relevant to the eight case histories but were, moreover, substantiated by a scrutiny of reports written on sixty other companies visited during the study. The classification is detailed in Figure 2.2, from which six levels of technology are designated as a suitable means of discriminating between innovative firms. It further follows from the classification that development associated with technology levels 1 and 2 will, in general, be evolutionary in character, technology level 3 will be mainly evolutionary but with some innovation, at level 4 it will be mainly innovative but partly evolutionary, while at the two upper levels 5 and 6 it will be wholly innovative.

Company No. 1 is of interest in that it is the only manufacturer of the eight which is not engaged in the production of engineering-based products. Its problem-solving activities are less demanding than those used in company No. 2, and, while its activities might merit publication of an article, it would be on grounds of general interest rather than on the nature of its technology. Development is clearly evolutionary and is carried out by craftsmen.

2.4 The Level of Technology as a Basic Determinant

The contrasting features between the positional and innovative companies Nos 1 and 2 and the innovative company No. 8 (described in section 2.3 above) are summarized in Figure 2.3 at levels of technology 1, 2, and 6. It remains to be verified whether these and the characteristics of the remaining five companies show an orderly passage from the lowest to the highest rating.

The characteristics of companies at technical levels 3, 4, and 6 are again derived from the case histories and are shown in Figure 2.3, columns II to IX. The picture is an orderly one, and confirms the potential value of the scale.

The distinguishing features between technology levels 1 and 2 are mainly reflected in the qualifications of staff needed to solve problems likely to be encountered in product design (see Figure 2.2). Should circumstances dictate that a company at level 1 be required to counter competition by undertaking defensive development for the first time, it would need to obtain staff capable of solving more difficult problems and would eventually move to level 2.

			Column		
I	II	III	IV	V	VI
Level of technology	R&D, D&D costs — Capital	Risk and Benefit pay-back Revenue period	Product range/ life/nos.	Sources of ideas	Recommended research development and design attitudes
1 – – – – 2	Low Low	Low Low	One basic product. Long life. Large production quantities	In-house and customer contacts	Defensive development and design in fields consonant with existing skills and resources
3 – – – – – – 4	Low/ Low/ medium medium Medium/ Medium/ high high	Low/ Low/ medium medium Medium Medium	Several products for a number of related markets. Medium life. Medium production quantities	In-house research, design and development staff in conjunction with marketing	Market-orientated and aggressive research, development and design using familiar techniques in current disciplines
5 – – – – 6	High High	High High	One up-market product per division or subsidiary. Long life. Small production quantities	In-house research staff, customers, and places of learning	Aggressive research in own and new disciplines

Figure 2.3   Level of technology — a determinant f

At levels of technology 3 and 4 the main differences stem from the emphasis given to evolutionary or innovative development and are reflected in financial, risk, and benefit ratios (columns II and III, Figure 2.3). The two pairs of companies at technical levels 3 and 4 differ from the remaining four companies at technical levels 1, 2, 5, and 6 in many important ways. The companies have a much larger portfolio of products, most of which are manufactured in smaller quantities than at technology levels 1 and 2, and larger than at levels 5 and 6. The operation of the businesses assumes a far higher degree of complexity than at levels 1 and 2, which is reflected in the manufacturing operations, the organization structure, and indeed in every aspect of management. Reference to case histories 3, 4, 5, and 6 also indicates that a further differentiating feature is the extent to which marketing, or sales, work closely with development staff. Not only is the collaboration more evident at level 4 but there is a tendency to adopt a more aggressive research strategy over a wider range of products. Should difficulty be experienced in assigning the appropriate level of technology, it will be helpful if an estimate can be obtained of the percentage of disposable funds allocated for

	Column		
	VIII	IX	X
ecessary company tributes	Recommended paths to growth	Main threat	Relevant case history company no.
gh production uantities at inimum costs. ood at distribution d selling	Find new home market segment and new export markets. Improve price and non-price factors relating to product/process. Move up to the next level of technology	Competitors' efficiency at producing and selling. Imports from developing countries	1 and 2
apacity to deal ith a wide spectrum activities. Good R&D, production, arketing, and lling	Improve sales by attention to non-price factors relating to product	Competitors achieve higher level of technology	3 and 4
	Evolve product		
	Radically improve product		5 and 6
sion and nfidence to invest the future. cellent at novating new oducts and selling	Consolidate world markets and maintain reputation for excellence	Commercial failure by reason of overstretched financial commitment. Failure to maintain world leadership	7
	Extend product range through radical improvements (Limited market)		
	Market additional unique products outside present range (Elastic market)		8

nodel for innovation in a manufacturing company

investment in technological change. If the figure exceeds the average for the relevant industrial sector the appropriate rating will be 4. The main threat to companies at technical levels 3 and 4 (column IX, Figure 2.3) stems from competitors improving the design and performance of their product range, whereas at technology levels 1 and 2 the threat arises more from non-technical factors; and at levels 5 and 6 the challenge is to avoid overstretching financial resources as a result of allocating heavy research expenditure to projects associated with a too-high risk.

Case histories at levels of technology 5 and 6 refer to innovative companies whose products have a reputation for excellence. Company No. 8 at technical level 6 is given the higher rating on the grounds that its scientists have collectively achieved an international rather than a national reputation. Company No. 7 has one outstanding product and is content to grow by using its research, design, and development expertise to make changes to the basic product which will facilitate entry into new home and export market segments. Reliance on one product is not a satisfactory long-term strategy, for it inevitably results in a weakening of innovative advances.

2.5 Manufacturers Who Use External Research, Design, and Development Resources

The acquisition of know-how from external sources, whether by licensing, consulting, or sub-contracting, is unlikely to distort the model at levels of technology 5 and 6. Products at these levels cannot be satisfactorily developed and manufactured without a thorough understanding of the pertinent scientific and technical principles, and, because the staff will necessarily be highly qualified, the intricacies of any new specialized disciplines will be quickly mastered.

At the two lowest levels of technology the products fall in the category of inventions, or even gadgets, based on well-established technical principles. If outside help is sought for product-development it is usually restricted to relatively uncomplicated design features of a functional and aesthetic nature. However, because manufacturing efficiency is crucial for this class of product, assistance may be sought with production processes, but this will only rarely affect the level of technology of the product.

Companies operating at levels of technology 3 and 4 might be expected to be anomalous, since they frequently manufacture a wide range of products for a number of different markets, and are engaged in both evolutionary and innovative development. However, should they seek outside help it is unlikely to affect more than a small proportion of their activities and will not therefore distort the model. Company No. 3 is a typical example of a company at technical level 3. Its management resources are directed to selling a range of products developed and manufactured in-house, selling licensed products which are assembled from purchased and manufactured components, and factoring a range of articles purchased from another manufacturer. All three operations are in accord with the activities shown in Figures 2.2 and 2.3.

A frequently observed difficulty for technology levels 3 and 4 may be seen in countries which are too small to support a sizeable and comprehensive industrial base, for they are inclined to operate overseas license agreements without the necessary technical support corresponding to the level of technology of the products. This has two inherent dangers: transfer of information from the licensee will rarely be sufficiently rapid to cope with local competition and, in the absence of staff with the necessary skill, unexpected manufacturing difficulties will be a continuing problem. Indeed, it is pertinent to recall that one reason for the outstanding Japanese industrial advance after World War II was their skill in selecting the most competent licensees and their determination to provide resources at the appropriate level of technology.

A common source of external expertise is through the purchase of component parts, the development of which has required a level of technology much in advance of that associated with the final product. An example of this apparent anomaly is the mass-produced car. Reference to Figure 2.2 would suggest a rating of 3, whereas Figure 2.3 suggests level 4.

The higher rating is a function of bought-in components and it is more appropriate to use lower figures. Specialist car manufacturers, such as Rolls Royce, also purchase many highly specialized components, but the calibre of their scientific and engineering staff and the nature of their residual research and development merits a rating of 6 or even 7.

Although a predominant share of a market may be divided among a few large manufacturers, sufficient segments will usually remain to support a number of small businesses. They typically operate over a local geographical area, have the advantage of low transport costs, and have the means to offer a competitive service. For these reasons, they market products which, while low in technical content, will be necessarily produced in small quantities. For the purpose of the model, these smaller companies may be regarded as an adjunct of their large rivals. They meet a need which has already been established and their products differ little from those marketed by their competitors. Their fortunes will be tied to that of the larger manufacturers, and once the latter lose their markets to an alternative technology, the smaller companies will need to respond rapidly by marketing a new product or go into liquidation. Smaller companies which depend upon orders from a small number of large manufacturers in one industry are likewise in a hazardous position. Their technology will normally be at levels 1 or 2, and Figure 2.3 indicates that, although they fit the model when regarded as an extension of their customers' activities, their shortcomings in columns V and VI mean that failure of their customers can only be countered by acquisition or by finding new markets through the use of external sources of development and design. Few companies with a long history of dependence succeed in finding new products from their own resources.

2.6 Strategies Appropriate to the Six Levels of Technology

Innovation is important at technical levels 4, 5, and 6, and the aim of companies operating at this level must primarily be to generate profits at a level consistent with funding radically improved products, knowing that major ones may result in a negative cash flow for at least five years. The strategic plan should seek to increase efficiency throughout the company, stimulate low-cost product innovation, and so provide funds to finance new ideas from which major developments will stem.

At technical levels 3 and 4 the company functions listed in Figure 2.1 will be intermediate between the positional and innovative modes and so be capable of dealing with a variety of evolutionary and innovative activities. For example, should there be, as is likely, a sizeable portfolio of new products and a concurrent need to improve manufacturing processes, considerable organizational and attitudinal flexibility will be required. Because a number of products will necessarily compete for available resources, higher management will need to institute formal methods to control and monitor progress. A necessary strategy will be to accept established procedures for developing complex projects and yet avoid burdening the less demanding innovations with too much control. One way of doing this is to assign the latter to lower levels of management and to encourage informal co-operation between *ad hoc* teams representing design, development, marketing, and sales.

An alternative strategy, which many companies have adopted to minimize the difficulties of disturbing stable organizations through the introduction of radical innovation, is to form new business venture-groups which are, in effect, loose appendages to the main organization. This strategy can be successful, but the incidence of failure is extremely high. Numerous reasons have been, and are, advanced for their failure, and although arguments continue there is sufficient evidence to enunciate two necessary criteria for success. The manager of a venture group must be told the nature and extent of resources which will be made available and, if possible, guaranteed over the first four or five years; and the conditions governing his appointment must be sufficiently potentially rewarding and yet, at the same time, sufficiently challenging to deter anyone who does not possess entrepreneurial qualities.

At a time when the rate of technical advance is quickening, companies at technical levels 3 or 4 may find that the only way to meet competition is by moving up-market to levels 5 or even 6. Not only will they need energy to surmount the inertia associated with a given level of technology but they will also require great determination to face the widespread change in practices and attitudes. A board will need to accept a higher degree of uncertainty and to be prepared to modify company functions. The organization will need to be more organic, the marketing more forward-looking, and the research and development more aggressive. It will be necessary to appoint higher-qualified scientists and technologists and more expenditure must be allocated to future activities.

Manufacturing operations are likely to become less of an obstacle to change than hitherto. The many new technologies, e.g. NC, CAD, CAM, and robotics, lessen the distinction between large-scale and small-scale production, since a factory may be subdivided into small autonomous units and yet be integrated by the use of mini-computers to give the advantages normally associated with mass-production.

At the two lowest levels of technology, and particularly for companies supplying consumer-durables, satisfactory profitability may be sustained over very long periods. However, should a competitor succeed in acquiring an ever-increasing share of the market it may be necessary to develop products at the next higher level of technology. Because level 3 is concerned more with evolution than innovation, the transition will be a gradual one, and the changes from a positional stance can be made without great difficulty. It follows that should a company at technical level 2 wish to meet competition by moving up-market through licensing, or acquisition, it should first do so at level 3. Only after satisfactory operation at level 3 should a move be made to the more innovative level 4.

At the two highest levels of technology a problem could arise if a company decided to produce a down-market product. It would be inexperienced in producing, controlling, and marketing large production quantities, and the preferred strategy would be acquisition or a merger.

2.7   Guidelines and the Six Levels of Technology

If the best choice of strategies for managing innovation depends upon the level of technology then it follows that it should be possible to decide which of the guidelines in Figures 1.2 to 1.7 are appropriate to particular strategies and hence to each level of technology.

Levels of technology 1 and 2 are concerned with evolution rather than innovation and this immediately suggests that many of the guidelines in Figures 1.4 to 1.6 will be inapplicable, since they are specifically aimed at innovative rather than evolutionary practice. Referring to Figure 1.4, only two of the eight need be retained, namely guidelines ii and viii. The remainder would require a degree of expertise only found at technical level 3 and above.

Guideline ii, Figure 1.6, is relevant at the lower as well as at the higher levels of technology. When a product is seen to behave in an unexpected way it is important to find out why. Is it attributable to the product being non-standard, to changes in the imposed stress patterns, or to a combination of both? The aim must be to isolate the variable or variables concerned, in the expectation that favourable modifications can be made which will improve the product. This can be a powerful way of achieving evolutionary development, but it is rarely used. Anomalies tend to pass unobserved because they are rejected by the logical processes of the brain.

The description of a product in terms of its elements and sub-elements (Figure 1.6, ii) can also be used to achieve improved products whether by small or large steps.[3] The number of times a product can be subdivided into constituent parts depends upon its initial complexity, but with four or five divisions the process is likely to reach the molecular level. With each successive subdivision of a product it becomes increasingly likely that science rather than technology is involved. In guidelines iii to vi the emphasis is on innovation and all three can be omitted. While the conflict situation with which guideline vii is concerned is most likely to stem from radical changes attributable to the innovation process, it can arise in any company and should be retained.

Figure 1.7 (innovation and design) is concerned to create an awareness of the particular demands which design makes on the management of innovation. The particular guidelines will, however, rarely (if ever) apply to technical levels 1 and 2 because the products are developed by craftsmen rather than by professionally qualified designers.

Companies at technical levels 5 and 6 employ scientists of distinction, and manufacture a small range of unique products. This being so, many of the guidelines in Figure 1.2, concerned as they are with an assessment of a

32

Fig.	Title of Guideline	Level of Technology		
		1 and 2	3 and 4	5 and 6
1.2	Innovation and the Company Board			iii, iv, and v
1.3	Innovation and Company Director			i, iii, and vii
1.4	Innovation and Company Environment	vii	vii	All
1.5	Organising Research, Development, and Design for Innovation	i, iii, iv, v, vi, and vii		
1.6	Innovation in Research Development	iii, iv, v, and vi		vi
1.7	Innovation and Design	All		ii, iv

Figure 2.4   Guidelines recommended for deletion from sets of guidelines in Figures 1.2 to 1.7

company's products against those of a competitor, are irrelevant. Accordingly, numbers iii, iv, and v can be eliminated. In Figure 1.3, i, ii, and vii, which are concerned with market share, selling, distribution, pricing, and constructing a product portfolio, will not require regular consideration at the two top levels of technology. Figure 1.4 (innovation and the company environment) is concerned to encourage, within the average company, a creative ethos. But as this is normally common to companies who sustain a reputation for excellence, these guidelines can be omitted.

Additionally, high-technology companies will have no need to establish venture groups (Figure 1.6, vi), or be adjured to appoint designers of the highest competence (Figure 1.7, ii), nor will they be required to take specific action to cultivate an awareness of current technologies and methodologies (Figure 1.7, iv).

If a number of guidelines can be omitted from the two lower levels of technology for the reason that staff is rarely sufficiently competent to carry out innovation, and guidelines can be disregarded at levels 5 and 6 on the grounds that staff is highly accomplished, it remains to examine whether arguments can be advanced to prune guidelines at levels of technology 3 and 4. Unfortunately this does not seem possible and, from the above discussions, it will be apparent that this is a consquence of the difficulties emanating from combining evolutionary and innovative practices. Companies operating at levels of technology 3 and 4 comprise the major proportion of manufacturing companies and are the most difficult to manage. Their manufacturing processes must be efficient and flexible, the boards must tread a delicate balance between financial prudence and the acceptance of risk, a management philosophy must be devised which has the advantages (but none of the disadvantages) of the positional and innovative modes, and competition is likely to be more severe than at the other four technical levels.

The concept of using the level of technology as a means of discriminating between companies has facilitated a rational pruning of guidelines at levels 1, 2, 5, and 6, and the recommended deletions are shown in Figure 2.4.

References

1. Nyström, H. (1979). *Creativity and Innovation*, Chichester, Wiley.
2. Langrish, J., Gibbons, M., Evans, W.G., Jevons, F.R. (1972). *Wealth from Knowledge: Studies of Innovation in Industry*, London, Macmillan.
3. Parker, R.C. (1972). 'On the Comprehensive Nature of R & D', *R & D Management*, **3**, No. 1 37.

CHAPTER 3

G.T. Culpitt & Son Limited
(Case History No. 1)

Inherited businesses are less likely to experience the same economic and social scene which gave them birth, and can, therefore, only be perpetuated by a Darwinian type of adaptive behaviour. This case history is an example of such a family business, which is encountering economic difficulties that stem from considerations largely outside the company's control. It is an exciting study for the student since, at the instigation of the company chairman, successful use has and is being made of current management techniques.

3.1 Introduction

Culpitts manufacture cake decorations, and, at their headquarters in Hatfield, they mount a most impressive display of their products. Their cake decorations meet the needs of every conceivable occasion, ranging from sugar-based edible flowers, motifs, and nursery figures to gold and silver lacquered ornaments, wedding sprays, bells, horseshoes, and many other novelties; stands, pillars, and cake-bases are produced in a wide variety of designs. All of their products are illustrated in a colour catalogue which is most attractive, as indeed are the products themselves. Figure 3.1 shows three typical decorations.

The company is currently under the control of two brothers, Peter and David Culpitt, who are chairman and managing director, respectively. Their father and grandparents began the business in about 1920; their grandmother was an artificial flower-maker to the millinery trade and, when orders fell short, she extended her artistic talent into floral bouquets for weddings and then cake decorations. With her husband and son they then started the present company from what was virtually a cottage industry. (Peter Culpitt's daughter is now in the business and his son is undergoing training with a view to joining the firm in the near future.)

In the beginnings of the company there were many competitors and German imports imposed a serious threat since 'they were not made of sugar and were not, at that time, very expensive'. However, 'Culpitt's share of the market had to be fought for and was greatly improved'. At the end of World War II it was decided to make a supreme effort to build up their production and stocks to satisfy the home market completely prior to a recovery of the German import trade; this policy was both courageous and successful. Using

34

Figure 3.1 Typical cake decorations

all available finance they took on many more staff and moved up to a new factory at Ashington in Northumberland, which was a development area. About this time a family friend began making small imitation Christmas trees for them near his home in Eastbourne, Sussex. Somewhat fortuitously, this small spur grew and extended into bigger premises and now, together with the Ashington factory, has a turnover in excess of £2m per annum.

There are obvious communication and transport penalties in having a business so widely split geographically, but Culpitts survive these without serious hardship, largely by carefully selecting the staff to manage their factories and by giving them a vested interest in the success of the company. This leaves the two senior partners free to concentrate on the more important tasks of policy-making and management, whilst the day-to-day production ripples are sorted out quite competently by the people responsible for them.

3.2 The Situation in 1978

Over the years Culpitts have expanded steadily. Their share of the home market, 60 per cent and 80 per cent of their business, is with wholesalers and cake-manufacturers. Today, they have a very dedicated board including company secretary, marketing, and production directors. (See Appendix I for company organization chart.) They have a staff of over 350, of whom half are non-productive. It is a successful, lively, enterprising company enjoying a pleasant atmosphere and good labour relations. Like many other small- to medium-sized businesses, however, Culpitts have not been without their problems and these have been particularly exacerbated by the adverse economic climate of the last decade. Many of Culpitt's operational problems stem from two particular characteristics of their type of business—it is seasonal and it is very labour-intensive.

Their dependence on manual labour in their factories causes the Culpitt board concern. The high minimum wages they have to pay—even for low-grade jobs—causes them to raise the price of their products to a level which endangers their survival. Furthermore, they feel extremely vulnerable to doctrinaire policies of present-day trade union control in labour-intensive factories such as theirs. Culpitts do have very good labour relations with their workforce; they achieve this by having the right people managing their factories. This enables them to maintain a contented pleasant atmosphere, to anticipate problems before they get too serious, and to delegate the authority to put things right.

Nevertheless, whatever plans the company might have for expansion do not include for any increase in their workforce. On the contrary, it is now their aim to replace manual labour with automated machine processes wherever possible. The chief obstacle to implementing this policy, however, is the high cost of production machinery, most of which is highly specialized and, with profits of only about 3–4 per cent of turnover, they just do not have the spare capital to invest in such plant to the scale they would wish.

It is well known that large quantities of small cut-price products are imported from the Far East, including plastic decorations, and should this trade expand its repercussions could be serious. As Peter Culpitt explained, 'We have always had competition from Hong Kong and Taiwan but, with rapidly rising labour costs in Britain, it is likely to become serious'. In an attempt to counteract this, the company has begun a small production factory in Mauritius, where labour is very much less expensive and where the Mauritian government offers considerable incentives to encourage foreign business development. The manager in charge is Mauritian and he is on the board of directors. Products and raw materials are despatched to the Mauritian section from the UK for painting and finishing by hand. Mauritian girls are extremely good at artistic work. Certain edible products are unsuitable for this treatment due to the hot, humid climate, but this would not prevent expansion of that section in other areas. At present, only twelve people are employed there. It will be more profitable with a larger workforce to offset export overheads.

They have a further problem in that 50 per cent of their business is done in three months of the year, from September to November, for the Christmas trade. This causes them severe problems with cash flow. Normally they have a credit balance for only a few months in each year (see Appendix VI). After paying off their overdraft and creditors, they are again overdrawn for several months. With this situation it is clearly important that they maintain strict budgetary control of their stock and order schedules and, to do so, they recently installed a computer which has proved invaluable. They have a particularly good operator on this, working in liaison with their company secretary.

The seasonal nature of their business is exacerbated by the inflationary situation in the UK. Three years ago, for example, they failed to anticipate the unprecedented rise in the cost of raw materials. As is the nature of their business, their stocks and orders to which they were committed had been arranged six months previously and they had pitched their prices too low. They could do little about it and they made a major loss that year. The bank refused to increase their overdraft and they were forced to seek help from a merchant bank, which requested 25 per cent of their equity before making the finance available to the company to cover their requirements. It is confidently believed that their computer-based budgetary control system will now prevent this type of crisis recurring.

Their North American business is covered by a full-time sales representative but 'by the time we've processed the orders he's secured, inflation has forced us to raise our prices and the customer doesn't want to know any more'. Perhaps it was not unexpected but, on being asked whether quality, delivery, or price was considered to be the most important factor, Peter Culpitt replied: 'Price was definitely first, followed by quality and then delivery.' As he explained: 'These kinds of problems are always with us. A business like ours cannot be run to a mathematical formula. We have to make decisions on what we should buy and produce, based on what we think we can sell at a price we think people will pay. But the balance is so delicate; a mistake of 2 per cent can cut our profit in half. One lorry-drivers' strike at the wrong time and our profits for the year are gone.'

So this was the position of Culpitt and Son Limited at the beginning of 1977—a family company which had enjoyed relative stability and security in the operation of its business for over fifty years being forced into changing its policies to stave off the threats and pressures of adverse economies and political climate. Basically the board knew what had to be done. Less-seasonal lines from new marketing areas would improve their cash-flow problems and automated processes would reduce their dependence on manual labour—this would go a long way towards stabilizing their prices and counteract the threat of cheap foreign imports. But how was this to be achieved? A single machine for automatically producing edible flower petals—even if it could be found—would cost over £30 000.

3.3 Manufacturing Operations in the Eastbourne Factory

The manufacturing processes in the factory are indeed very labour-intensive, and many of the operations rely on the skill of the operator: for example, flower petals were hand-made individually by squeezing out the appropriate amount of material from a bag. Considerable initiative had been shown in the construction of semi-automatic plant from standard components. It transpired that much of this equipment had been designed and adapted by Peter Culpitt himself, at home with his own lathe, milling machine, and other tools. He refused to call himself an engineer, yet the skill and ingenuity evident in his creations were far superior to those of many people who claim that title. Silk-screen printing was used to print designs and messages on edible wafers. Orders for these ranged from 10 000 to 150 000 at a time and the production rate was around 2500 per day. This method of silk-screen printing on edible products had been developed by the Culpitt brothers and was one of several patents held by them to protect their processes. One product which uses the printing is a decorated waferette. In another room, circular hand-made communion wafers were made from starch on which was embossed a small cross. An edible dye was applied by a silk-screen process. They were then cut and stored automatically. There was, incidentally, a 50 per cent wastage of material cut from around the disc in this process which amounted to £150 per month. The manufacturing of tiny Christmas trees was another standard line. These were made largely by hand: small lengths of green Mexican fibres were hand-placed between pairs of wires which were twisted to form the radial 'branches'; these were trimmed to a conical shape and given a snow-covering by dipping in a paste solution and spinning by hand so that only the tips of the branches were coated with little white blobs.

In another section, edible spheres were made from sheets of a starch solution pressed between heated male and female platens, which were machined with hemispherical undulations. This machine operated on a 20-s cycle and on removal from the press the resulting hemispheres were cut by hand from the pressed sheets; another operator then stuck two together to form hollow spheres; these formed the head and body of small decorated figures for birthday cakes. The painting of all their cake decorations was done separately by hand. It was indeed evident that, with so much manual work, the company was heavily dependent on a stable and consistent workforce. The seasonal nature of the employment ran counter to this since, outside the festive season, much of the staff would either be stock-piling or under-employed—either way, cash-flow problems would clearly be paramount during the summer months.

There was one new project that the company was in the process of developing which had the potential to even out their business cycle during the lean periods. This was an order for small imitation trees to be used in United States' flight-simulator models measuring 100 ft × 40 ft. These trees had to look realistic when magnified by television cameras and about 17 million were needed. During the visit the manufacturing techniques for this order were far from complete and the company was behind schedule. The apparatus for making the trees from cork and wire was quite ingenious and two people were working on this at the time of the visit. The device was largely home-made—again to a design conceived by Peter Culpitt.

This project, and possibly others like it, seemed well within the capabilities and expertise of the company and, as stated above, it might go a long way towards solving their irregular order schedules. It seemed rather surprising, therefore, that more effort was not being made to push ahead with this opportunity or to investigate similar market areas. The order for the trees came in the first place from the Americans but, despite its obvious importance, there was no mention of a marketing enquiry having been made to other airlines, filming organizations, scenic model-makers, etc.

During the visit there was a problem with a machine for producing starch-based wafers. This was a gas-heated press—a large, fairly old, Baker–Perkins model, bought second-hand for £750 and which had been cleverly adapted to perform its current role. Inside, a chain of eighteen platens were sequentially opened, fed with a mix, and closed; they passed into an oven and were 'pressed' together by steam with a force of 4 tons. The machine was producing an inconsistent and unsatisfactory product, and a cure was being sought by a number of senior staff and workers on an *ad hoc* basis. The corporate spirit with which the problem was being tackled clearly exemplified the very good labour relations enjoyed by this family business but it really was uneconomic in staff deployment. The cut-and-dry technique adopted, although requiring a fair degree of skill, lacked any real methodology or system.

Process problems of this nature are particularly difficult, since the ultimate solution would be likely to involve alterations to a number of factors, e.g. temperature, viscosity, time, pressure, etc. Indeed, the recommended approach is to use a statistical planning technique designed to measure the combined effect of sets of possible interacting variables. However, it was perhaps inevitable that the management kept hoping that the next experiment would succeed and did not seek help with an alternative approach. In fact, eight months elapsed before this machine fault was determined and rectified—eventually, by chromium-plating the platens and applying an anti-sticking agent. The delay clearly cost an estimated £30 000, to which should be added the lost opportunity costs.

3.4 Conclusions After Factory Visits

Notwithstanding the difficulties associated with seasonal production in a labour-intensive industry, Culpitts were seen to be a reasonably successful family business, enjoying pleasant and cordial staff relations. They were coping with their problems and showing a profit, albeit a small one. The company was keeping its head above water largely due to the enthusiasm and initiative of the senior partners, for whom the staff had considerable respect. Nevertheless, they did have cause for concern, and their current state of affairs could very quickly change for the worse if something was not done in time.

First, Culpitts appeared too reliant on traditional outlets and were becoming increasingly vulnerable to overseas competitors. To protect themselves against this threat, new market opportunities had to be sought and a planned marketing policy formulated and put into practice quickly. Second, an attempt had to be made to even out their cash-flow situation so that the new market outlets should take account of this.

Under the inspiration of the chairman there was considerable evidence of ingenuity in the construction of semi-automatic equipment in the factory but innovation was really needed on a much broader front to embrace entire work-processes. Furthermore, these processes would need to be designed in such a way as to be less dependent on manual labour—particularly if they could not be fully utilized all the year round. This desire for fully automatic processes was, in fact, the objective which prompted Peter Culpitt to seek outside advice in the first place, but their current methods of operation conflicted with this form of development; there was little evidence of planning which operations would be best automated having regard to profit capital, market importance, etc.

During many years of evolution and development of a factory it is inevitable that much of the development occurs fortuitously and in several unco-ordinated directions. This indicates a healthy determination to follow up particular ideas and projects. To give one example, one order from a large supermarket chain amounted to 20 per cent of the business and its many 'specials' made high production efficiency almost impossible to attain. Periodically, however, it becomes necessary to stand back and take an overall look at a situation, with the object, for example, of reorganizing the layout more efficiently. It was felt that the Culpitt factory had reached such a stage and the time had arrived when a more purposeful direction might be given to the innovation of work process—especially in the establishment of criteria for development in relation to in-house and external sources of expertise.

3.5 Further Action

It was believed that the firm would benefit considerably from the expertise of external planning consultants to enable them to rationalize their production and marketing methods. They were given details of the Manufacturing Advisory Service, offered by The Production Engineering Research Association (PERA), which has all the facilities and expertise to provide information and assistance in all areas of manufacture, materials, marketing, financial control, etc. PERA is a non-profit research and development consultancy, training, and information centre set up by British Industry and the government in 1946 to provide individual firms with help to overcome specific production and management problems.

Although it was not the function of the 'Management of Innovation' project team to act as planning consultants it was felt that in this particular case, when the company might be at a crossroads, there would be benefit to both the team and the company in undertaking a small work programme to help steer the firm towards better planning and regionalization of its operations. Given the goodwill and co-operation of the chairman, an effort in this direction could increase their profit margin significantly. To continue as they were might easily result in their being forced out of business if overseas competition based on cheap labour were to undermine their traditional lines.

3.6 Synectics Session

To identify new market opportunities which would be less vulnerable to overseas competition and offset the seasonal fluctuations of traditional business the project team suggested the use of the synectics creative problem-solving technique (see Appendix II). A group of seven was selected and comprised Peter Culpitt, Dr Parker, the plant director from Ashington, and four other staff members representing different functions and levels of hierarchy. An external specialist conducted the sessions.

After a brief explanation of the principles of synectics, including suspension of judgement, use of fantasy and apparently irrelevant material, constructive evaluation, etc., the group addressed the question 'What products can we develop using existing technology?' Initially they generated twenty-five goal/wish statements, covering a wide range of possible approaches, wishes, embryonic ideas, etc. (see Appendix III). Three of these were selected for further development:

(1) A new kind of sweet using Culpitt's printing technology;
(2) Mail-order sales of complete cake-decorating packs; and
(3) Metalized plastic products.

Four specific variants of the first concept were developed—sugar chocolate and printed wafer petals. Each idea was given a percentage rating by the participants for its newness, appeal, feasibility, and probability of success. The next session was devoted to the concept of mail-order cake-decorations. The concept was rated as before and scored high on all counts. Possible problems were identified and ideas generated on selling methods, products to be offered, designs, etc. The session ended with a listing of the action steps to be taken to pursue the idea. The third session dealt with metallized plastic products: a list of products for metallizing was generated and each participant selected their preferred idea from the list.

After a lunch break the group reassembled for one further session involving a high level of speculation with the object of developing a novel idea. The group generated a further twenty-seven goal/wish statements, with a much higher level of fantasy (lower practicality) than in the first session. They then used imaging techniques to generate fantasy material and developed intentionally absurd ideas. By a process of triggering and association a highly novel concept emerged: 'A wafer containing the ingredients of medical pills cut into strips, with symbols printed on, for underdeveloped countries.' The idea was considered by the group to be highly novel, appealing, and feasible, and the chairman decided to put in a patent application and talk to a pharmaceutical consultant about it.

Subsequent to the synectics session (which took about 4½ hours) the planning director from Ashington attended the five-day synectics course with a view to using the techniques in the company on a regular basis. Some two months later Peter Culpitt held a 'think-tank' session concerned with their corporate planning exercise and explored the company objectives, corporate image, current position, threats, strengths, and weaknesses (see Appendix IV). Additionally, they had five ideas for new products, and it was interesting to note that, from those five ideas, three were identical to those obtained in the seven suggestions from the synectics session, although the two groups only had two members in common.

The current position of the main suggestions is that the research and development team has been asked to design a machine to produce a sugar-wafer sprinkle. Peter Culpitt commented that 'No other person in Europe has such a machine, so a large market could develop of a non-seasonal nature'. It was, however, decided to postpone direct mail-order selling because of staff wage problems. Again, while metallizing plastic was believed to have a future, it was deferred because of lack of spare production capacity and of finance to purchase machinery. Chocolate-containing projects had to be abandoned since the company had insufficient expertise with this material. A new product in new market segments, mainly drugs in wafers, is being patented and considered by a drug company.

3.7 Corporate Planning

The success of the synectics session appeared to help the company accelerate its move towards a more progressive management of their business. Perhaps for the first time, the board started to take a really critical look at their situation. They drew up a corporate plan which was based on an appraisal of their expertise, their strengths, and their weaknesses. They evaluated the threats and risks to which they were exposed and they considered alternatives to overcoming these. They stated their objectives carefully and realistically and set specific targets to be attained within a given time-scale. In addition, they have made 'brainstorming' sessions a regular feature of their company policy.

Culpitt's new strategy for forward planning involved four major policy changes:

(1) They made a critical assessment of all sections of their production department and used the simple criterion of profitability to decide which plant should be automated, semi-automated, or dropped. This, of course, involved accurate forecasting of market potential, capital outlay, and labour-savings.
(2) They would make further use of external expertise, manufacturing advising services, etc.
(3) They intensified their efforts to find new markets. One idea that stemmed from the synectics session was direct selling of their products by mail-order. They carried out market research to determine the demand for communion wafers. (This was done for them by two students from Buckingham Management College.) To Peter Culpitt's surprise here was a considerable market, possibly worth £35 000 in 1979 which, together with the model trees, could help to even-out their cash flow. Greater efforts were also to be made to increase their sales abroad. Fifteen per cent of their production was currently exported, with Canada being the best overseas market, but it was believed that this could be increased significantly. A major asset of the company was its highly motivated sales force—two in Europe, one in Scandinavia, and one in North America. It is of interest to note that the European market is restricted because several countries have no tradition of decorating cakes. New markets in Canada and the USA were thought to have considerable potential, and these would be served by the Mauritian factory.
(4) Research and development of new products would be made a key factor in their future strategy. To this end, a small development section was set up at Radlett near the company headquarters. Only two engineers were initially employed there, one aged thirty-eight the other twenty-five, but both had served apprenticeships and both had obtained their City and Guilds Certificate. Additionally, a development engineer was appointed to the parent factory at Ashington, whose principal duties were to adapt and improve machines to suit the Culpitt processes. The first project to be worked on at Radlett was an automatic tree-making machine for the American flight simulation model based on Peter Culpitt's original idea but better suited to production in terms of safety and reliability.

3.8 The Current Situation and Change

In the current inflationary climate things are still far from easy for the company. The reorganization of the factory has to be undertaken without disrupting current production schedules and there is little spare capital for this development. Profits are only 3 per cent of turnover and are daily eroded by price-increases in raw materials, mounting administration costs, and ever-increasing wage demands. In October 1978 the company's wages bill was increased by 22 per cent and the mark-up in profits was not sufficient to redress this higher expenditure; there seemed little prospect of improvement during 1979. The attitude of the TUC to our economy gives little incentive to growth and, to survive and prosper in such a climate, Culpitts are more than ever determined to reduce their workforce. The projections made in their corporate strategy and the planned automation of machinery were all aimed at increasing production and profits with fewer staff and, until this situation was achieved, the hardships would have to be borne without any real net increase in individual incomes. In the future, more consideration is to be given to licensing as a means of evening-out cash flow; indeed, this possibility has become part of their current thinking and discussions are currently being held with one British manufacturer.

Undoubtedly, Culpitt's new strategy contained a large element of innovation and it required considerable courage and dedication to carry it out. This was a healthy attitude, which had all the hallmarks of success for the future prosperity of the company. But perhaps the most innovative and courageous move of all was that made by Peter Culpitt of his own free will—to recognize the shortcomings of his own organization and to do something about it before it was too late; given the will, all things are possible.

Since completing the body of this report, Peter Culpitt has written: 'I would just add that, out of all the multiplicity of ideas, we only needed two or three to work out successfully to considerably change and improve the company's financial position and, in this connection, the effort put into the model trees for flight simulators has just resulted in a £75 000 order being received which, in itself, will be extremely profitable due to the fact that these items are manufactured completely automatically from start to finish as a result of producing one machine specifically to do this job. Several of the other new items are in the pipeline, the main one being the production of a machine to produce sugar wafer sprinkle—so here again a large market could develop of a non-seasonal nature.'

Appendix I Organisation Chart

Board of Directors:

P.J. Culpitt	Executive chairman
D.A. Culpitt	Managing director
E.J. Ford	Company secretary
D.C. Cooper	Marketing director
E.E. McCaw	Production director
Mrs C.K. Culpitt	

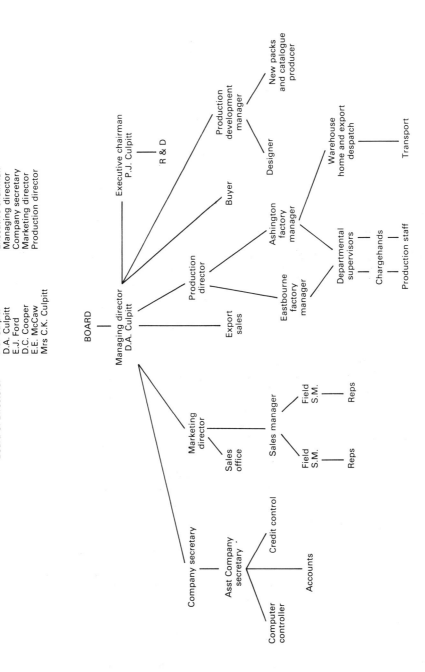

Appendix II   A Synectics Group

A synectics group comprises six to eight people; it is structured and the participants are trained to behave in a disciplined way. The chief role is that of the client with a problem, whose solution he has authority to implement.

A leader is chosen whose duty it is to steer the group in a prescribed way and he may not contribute to the content of the business. Three to five participants have the sole function of helping the client to solve his problem. A final important role is that of an observer, trained in synectics, who will watch the session on a videotape screen in an adjoining room and note everything of significance. At the end of a session he leads a discussion, replaying those sections which illustrate the three best points and the worst one. Most frequently, the latter occurs when a leader faces a new, unexpected situation and the group then tries to formulate a procedure to deal with any similar recurrence.

At the outset, the client describes his problem, indicates measures he has taken, and suggests the area in which he expects help. If his description is clear, questions are unnecessary. They are, in any case, discouraged as they are too often advanced to prove an idea unacceptable or show that the questioner is, in some way, superior. During this period participants write down their ideas to avoid being preoccupied with thoughts which seemed particularly relevant at the time.

Concepts are then suggested which, if realizable, might lead to a solution. These are expressed tersely and are written on the blackboard by the leader; therefore no-one feels that his contribution is lost or deprecated.

To engender a positive atmosphere concepts are expressed as a 'wish', and begin with the words 'How to...'. Since over fifty may emerge in a fast and stimulating stream, their source quickly becomes anonymous. At this stage there is no evaluation and therefore no criticism, and both leader and client encourage an atmosphere of goodwill.

Concepts cease when the client selects one which appears acceptable, possibly at an emotional level, and explains his choice. Then the leader seeks one idea from the participants which must fulfil the expressed wish. To ensure that he understands it this idea is paraphrased by the client, who then gives three reasons why he thinks it is good and outlines one reservation, should he have one. By concentrating on the good points he protects the author of the idea and encourages shy, imaginative participants, whom criticism might discourage. The leader next calls for an idea to meet the expressed objection, and the process is repeated until the client has no further doubts.

At the leader's discretion participants may be invited to enlarge on an idea before the client says why he likes it. This procedure may rapidly produce a solution, although one is not agreed till the client declares that it is new and feasible and states how he proposes to implement it.

Should a leader judge the participants' ideas as pedestrian, he will halt the session, select a word from the blackboard, and invite work-associations.

After a few minutes, he will choose an unusual word and ask the group to invent an original story, full of action, which is based on fantasy. At an opportune moment he will demand a solution in the fantasy world created. As new and exciting analogies generate fresh insights it only remains to adapt this solution to reality. This is called an 'excursion' or speculative session.

Appendix III   Synectics Session — Eastbourne 7 May 1978

Those present:

P.J.C.
E.E.M.
D.T.B.
A.I.
J.S.
J.McI.
Dr Parker
Vincent Nolan

Session 1

What products can we develop using existing technology?

1. How to make artificial jewellery for television and theatre in India, Africa, and the tropics.
2. How to cover events in different countries.
3. How to make photo frames in varying sizes with metallized finish.
4. Metallized products for Christmas tree decorations, paper and foil bands as Christmas and party streamers.
5. How to make cheap children's sweets with sherbet, etc.
6. How to have something pornographic.
7. How to go for teenage market.
8. How to make dishes to fit an oasis for flower-arranging.
9. How to have metallized plastic flower pots.
10. How to form cake-decorating clubs.
11. How to have cake-decorating parties like Tupperware.
12. How to teach cake-decorating in schools and colleges.
13. How to sponsor bright pupils in bakery schools.
14. How to teach people to use our products.
15. How to make piping tubes for use in cake-decorating.
16. How to provide complete cake-decorating packs including pillars, etc., through mail-order medium, including recipes, etc.
17. How to combine a new sort of sweet with printing technology, for example — *Star Wars* on a sugar sucker on a stick.

18. How to be more flexible with sweets.
19. How to respond quickly to current crazes and fads.
20. How to have printed ice-cream wafer.
21. How to print and wrap ice-cream wafers to take home, with child's name on.
22. How to make demountable icing top, i.e. preformed icing top.
23. How to make dissolvable confetti.
24. How to have small vanity mirrors for women's handbags.
25. How to make mirrors.

Start, concept and builds

A sweet, perhaps like sugarettes but printed on hard sugar. How to produce, on a machine, little circles and shapes. Some sort of sugar on wafer could be interlaced with chocolate. Design could change as you want, though it could be three layers. Tie this in with *Star Wars*, World Cup, and similar events and perhaps chocolate-printed. Pornographic wafer chocolate biscuit in the shape of a naked lady, perhaps on a stick. Rockets, necklaces, games, wafer chocolate printed on a stick, shaped letter in sugar in a ball of chocolate. The idea being that if you suck through the chocolate and get the correct letter, you get a prize.

Possible solutions

(1) Sugar chocolate sandwich, each layer with different design could be used as a game.
(2) Sugar-moulded items looking like *Star Wars*, etc.
(3) Sugar shapes rolled in chocolate, possibly strung together as beads.
(4) Combination of wafer petals with something in, with a printed letter or face on it.

Solution I — Group evaluation

It uses existing techniques, it has a big market for children, we cash in on the current crazes. We know we can do most of the items. Straightforward raw-material supplies; games could be promotional sales point. If it catches on the price is not too important.

Solution I — Negatives

How to get chocolate technology, experts, and conditions. How to improve packaging to deal with overseas conditions or deal with returned goods. How to satisfy food regulations, how to cut process to make it simple, how to make it completely different, how to make it unique, how to break into completely different sales lines. How to have different sales team to serve confectioners.

Sugar chocolate sandwich

Name	New	Appeal	Feasibility	Probability of success
Peter	50	60	50	50
Janet	80	70	60	50
John	60	70	50	50
Mac	90	80	40	10
Dave	50	60	40	40
Alan	60	20	20	20
Charles		70		

Moulds

Peter	50	70	70	60
Janet	60	70	70	70
John	70	70	70	60
Mac	60	80	90	80
Dave	50	50	40	40
Alan	80	60	80	40
Charles		60		

Sugar shapes rolled in chocolate

Peter	50	50	50	40
Janet	60	70	60	50
John	50	50	50	40
Mac	80	80	50	40
Dave	60	60	40	40
Alan	80	60	20	20
Charles		40		

Combination of wafer petals with something in

Peter	40	60	75	50
Janet	40	80	80	60
John	40	50	70	70
Mac	40	80	80	70
Dave	70	60	80	80
Alan	90	90	90	80
Charles		60		

Session 2

Mail-order cake-decorations

Sell cake-decorations through mail-order business, including decorating material, rolling pins, tins, piping tubes sold to housewife through mail-order business.

Form cake-decorating parties and clubs. Produce catalogue containing all the above items, also recipes. Complete availability of cake-decorations and articles to make cakes. If it goes on top, round, or underneath a cake we should sell it. We have to decide whether the catalogue should be sold or given away.

Name	New	Appeal	Feasibility	Probability of success
Mac	80	90	90	90
Dave	80	80	80	80
Alan	100	100	100	100
John	80	80	80	90
Janet	90	90	80	80
Peter	60	80	90	80
Charles	90	100	90	80

Possible problems:

How to avoid running out of stock, extra postage.

How to get catalogues sufficiently appealing to people who want to buy the product.

How to create designs the housewife could follow.

How to prevent interference with existing production.

How to get to young housewives.

Selling methods

Cake-decorating clubs, cake-decorating parties, cake-decorating classes. Teaching and instruction in schools and colleges, exhibitions — 'Ideal Home' for example. TV advertising, lecture to Women's Institutes, etc. Door-to-door canvassing, market surveys, free gifts to newly married women, giving prizes to training colleges, e.g. present the Culpitt Cup. Send out display van, coaches, or trains, advertising in magazines, mail shops, tie something in with promotions put out by other bodies such as the Milk Marketing Board, Tate and Lyle, or some other bakery or bakers.

Ideas

Use existing nucleus of cake-decorators. Use night schools, tape cassettes, recipes, and instructions. Use personalities such as Jimmy Young to promote ideas and sales.

Products

Silver and golden wedding decorations. Develop a theme similar to the Post Office theme 'Make someone happy' by telephoning them. Dial-a-recipe. Use product to celebrate personal events and successes like passing the driving test, in the same way that people sell get-well cards, birthday cards, etc. Silk-screen printing on sugar and wafer; print the message on a cake frill band, print on sugar paste. Make a limited number of shapes which can be made into a wide range of designs.

Stock problems

We need finance and a space for additional stock. Need for greater distribution, greater turnover, stricter housekeeping.

Designs

Need to develop designs for all-year events, such as Mother's Day, Father's Day. Need to be able to produce small batches more frequently.

Action/Steps	Action
Get source of supply for peripheral items like cake tins, rolling pins, piping tubes, etc.	M.G.
Finance requirement	P.J.C.
New team	D.A.C.
Find out who can come up with recipes. We need a celebrity, Jimmy Young, for example	New production manager
Present ideas to think-tank at Hatfield	P.J.C.
Present Gill Mentor with this idea	D.C.C.

Session 3

Metallized plastic	Preferred ideas
Alan	Photo frames
Dave	Christmas-tree decorations, bottle brushes
Mac	Teenage market
John	Paper chains
Janet	General decorations, party streamers both metallized and paper

| Peter | We need a gimmick in metallized plastic to sell to African and Asian communities relating to their social events, their religions and celebrations, also metallized festival decorations. |

Products for metallizing

Plastic model kits, metallized items put in bags of cornflakes, latest crazes, gold and silver discs to hang on chains round the neck. Tie up with Kellogg. Metallized pop star names. Tie in with promotions, metallized name plates, metallized visiting cards, metallized plant and flower pots, table vases, metallized place mats. Metallized plastic cake-boards, plastic metallized wedding cake-boxes with metallized tag. Half-reflecting spectacles. Metallized knives and forks metallized in gold, handles only. A range of metallized cards. How to get over the problem of wear. Chrome plastic things, teenager could put together, puzzles, craft kits, rockets, jewellery, pop-star names, and jewellery kits.

Session 4

Drugs, Wafer

1.30 p.m. Speculative session. (Here the aim is to 'trigger off' associated ideas.)

1. How to press a button to automate the cake-fill department.
2. How to have complete automation.
3. How to completely automate waferettes.
4. How to employ a load of monkeys.
5. How to double the wages with half the staff.
6. How to get a communion wafer that gives you a religious experience.
7. How to have coloured communion wafer, or flavoured.
8. How to have additives in communion wafer.
9. How to have Christmas-tree decoration that sings happy birthday.
10. How to have wedding albums that play the 'Wedding March'.
11. How to develop reflective strips for motor cars and clothing.
12. How to have cake-decorations that are fly-papers.
13. How to produce sequins for ballet dresses from metallized plastic.
14. How to have six- or seven-sided metallic revolving thing to titillate.
15. How to have metallized display trays and Christmas trees to replace parsley to contrast with meat.
16. How to cover every cake in Liverpool/England with our products.
17. How to make our own candles.
18. How to cover waferettes with wax.
19. How to cover plain waferettes and print crucifix on with warm iron.
20. How to emboss waferettes.

21. How to have plastic records and metallize them.
22. How to make them play a tune afterwards.
23. How to use metallized process to take a copy of something.
24. How to use embossed wafer with flavouring and colouring.
25. How to invent a method to stop people smoking.
26. How to have a non tooth decaying sweet.

Concept

Develop a wafer paper with ingredients of pills cut into strips with symbols printed on for underdeveloped countries.

Name	*New*	*Appeal*	*Feasibility*	*Probability of success*
Peter	75	75	80	30
Janet	80	90	80	70
John	80	90	80	80
Mac	80	80	60	50
Dave	90	60	75	70
Alan	80	60	80	70

Positives

We are experts in wafer, we are better able to do it than anyone else. We do not necessarily have to make ourselves. Could license process, easy to pack, might lead to more use of wafer, it is non-seasonal. It has worldwide sales. Considerably easier for people who cannot take pills as it goes soft and soggy. Easier to store in hospitals. Dosage regulations can be printed. Identity of drug can be printed. It is more safe and it is a way to induce children to take medicine.

Negatives

How to design a childproof pack to put them in. How to obtain medical know-how. How to obtain specific medical advice to know whether drugs could be heated. There is a need for safety precautions in productions. High cost of pharmaceutical standards. How to meet with worldwide demand. There is a need for completely separate production unit. Need for fully automated production.

Next action — Peter

Slap a patent application in. Get pharmaceutical consultant. Bring up at next think-tank. Idea wafer-sweeteners, saccharine, drug could be sprayed on or screen printed on afterwards, or microdot, cap-it like toy pistol caps, percussion caps. Drug could be included in ink medium.

Appendix IV   Company Brainstorming Session as Adjunct to Corporate Plan
Opportunities in General Terms

Our extensive knowledge of British and world markets, together with a variety of processes, should enable us to find a specialized 'window' into markets.

Our extensive past experience should enable us to be successful in new projects.

We can obtain some government finance for R & D projects.

Ashington capacity and administrative capacity.

Wafer-growth possible.

Problem   How to:

Achieve a profit of £350 000 on an expanded turnover by 1982, preferably using existing manufacturing UK space — profitable — maximizing strengths and minimizing weaknesses, especially cash flow. (See Appendices V and VI for balance sheet and budget analysis.)
Minimum range discussion potential of sales of £100 000 in 5 years.

Brainstorming Session

1. A   Take over another company allied to our manufacturing techniques. Established idea.
2. D   Start up mail-order business.
3. B   Expand retail sales.
4. D   Sell goods on a party-plan basis.
5. B   Mould our own plastics and expand.
6. D   Start bakery training schools.
7. E   Cups to bakery students.
8. B   Make more of our own raw materials.
9. G   Cheaper substrate for drums.
10. G   Look at range of products — eliminate poor profit lines.
11. G   Drop to 400 lines of 50%.
12. A   Take over Falcon Games.
13. G   Sack all salesmen.
14. G   Sell only to wholesale trade.
15. G   Reduce units to two.
16. C   Slant new products to maximize what we can get from government.
17. G   Use computer more.
18. C   Sugar or wafer sprinkle.

19. B Trees — model railways.
20. B Summer season — ice cream — sweets.
21. C Allied items — tableware.
22. B Sell to hotels.
23. B More to Northern Ireland.
24. B Create a US and Canada sales base.
25. D Run tubes and tins, etc.
26. C Into big market of bakery machinery.
27. F Sell licensing technology.
28. G Adhere to expense budgets.
29. A Take over Creeds.
30. G Sell and lease-back Eastbourne.
31. C Obtain manufacturing techniques from overseas. Joint companies, etc.
32. G Mechanize manufacturing processes.
33. G Import any product where labour is 50% more than cost price.
34. A Take over competitors.
35. A Take over competitors' positive cash flow.
36. G Eliminate short runs.
37. C Decorate and sell cakes.
38. C Major cake manufacturers — view to decorate their cakes.
39. E Sell corporate image to housewife.
40. D Nationwide kids' parties.
41. B Yogurt and mousse market.
42. E Co-operative advertising.
43. E TV advertisements.
44. B To Kelloggs — give-aways — promotional.
45. C Wafer-cutters and products — gingerbread.
46. B Expand into metallizing business.
47. B Communion wafers.
48. B Flavour of the month — designs on.
49. B Impregnation of wafer.
50. B Supply florist sundriesmen.
51. C Good at production overseas.
52. A Buy Hong Kong company.
53. C Make own candles.

A Take over — 1, 12, 29, 34, 35, 52
B Expand existing facilities — 3, 5, 8, 19, 20, 22, 23, 24, 41, 44, 46, 47, 48, 49, 50
C New product items — 16, 18, 21, 26, 31, 37, 38, 45, 51, 53
D Mail order — 2, 4, 6, 25, 40
E Promotional — 7, 39, 42, 43
F Licensing — 27
G Savings — 9, 10, 11, 13, 14, 15, 17, 28, 30, 32, 33, 36

Main Products Chosen to Investigate New-product Assessment

	Sugar wafer sprinkle	Mould plastics	Overseas techniques licensing manuf. vermicelli	Decorated cakes	Gingerbread men wafers	Direct-mail selling	Wafers for ice cream and mousse	Sweets	Comm. wafers	Metallizing
New	25	8	3	0	21	12	24	11	26	14
Appeal	18	21	2	−5	22	20	16	14	27	21
Feasible	17	11	10	−3	23	11	11	−2	30	11
Probability of success	14	11	−4	−11	18	10	9	−1	22	10
Total score of all those present	+74	+51	+11	−19	+84	+53	+60	+22	+105	+56

Scored by: Excellent   =   +3
              Good   =   +2
              Fair   =   +1

              Poor   =   −1
              Bad   =   −2
              Dangerous   =   −3

Savings — deal with.

Mechanize.

Promotional.

Get facts and report back.

Feasibility study — time-scale on achievements short and long.

Write up all papers.

Appendix V Balance Sheet G.T. Culpitt & Son Limited

	Note	1978 £	£	1977 £	£
Employment of Funds					
Fixed Assets	10		182 301		164 801
Current Assets					
Stock	1(c)	368 785		339 739	
Debtors (less provisions)		474 541		382 722	
Cash at bank and in hand		18 312		17 420	
		861 638		739 881	
Current Liabilities					
Creditors		440 733		383 537	
Bank overdrafts	11	30 628		25 422	
Current taxation		8 746		3 312	
		480 107		412 271	
		381 531		327 610	
Corporation tax payable on or after 1 January 1980	12	7 000		8 000	
			374 531		319 610
			556 832		484 411
Funds Employed					
Share Capital	5		46 812		46 812
Reserves	6		370 439		297 593
Subsidiary Companies	7		(174)		10 240
Convertible Loan Stock	8		50 000		50 000
Long-term Liabilities	9		960		960
Investment Grant	1(f)		3 948		4 520
Deferred Taxation	1(d)		84 847		74 286
			556 832		484 411

P.J. Culpitt } Directors
D.A. Culpitt }

A

Appendix VI Budget Analysis 1979

Budget: 1979 Account No.:: Description: Summary 13 November 1978

	Jan	Feb	Mar	Apr	May	Jun	Jul	Aug	Sep	Oct	Nov	Dec	Total
Sales	155 000	181 000	195 000	164 000	162 000	200 000	254 000	244 000	320 000	373 000	329 000	123 000	2 700 000
Cost of Sales													
Purchases													
Ashington	25 800	45 400	55 300	56 800	59 000	54 900	47 500	43 200	54 500	52 700	44 700	38 200	578 000
Eastbourne	9 500	10 400	10 900	11 900	9 000	8 800	9 200	9 900	9 800	11 000	9 100	7 500	117 000
Total Purchases	35 300	55 800	66 200	68 700	68 000	63 700	56 700	53 100	64 300	63 700	53 800	45 700	695 000
Direct Labour													
Ashington	19 399	27 445	34 875	28 331	30 497	29 665	24 214	21 812	30 562	31 529	34 212	29 666	342 207
Eastbourne	14 357	14 357	17 511	14 357	14 357	17 514	13 847	13 847	16 873	16 269	16 269	19 894	189 452
Total Direct Labour	33 756	41 802	52 386	42 688	44 854	47 179	38 061	35 659	47 435	47 798	50 481	49 560	531 659
Stock Adjustment	1 469	(15 247)	(29 861)	(36 768)	(39 144)	(19 879)	20 809	22 261	33 865	58 217	45 414	(39 295)	1 841
Total Cost of Sales	70 525	82 355	88 725	74 620	73 710	91 000	115 570	111 020	145 600	169 715	149 695	55 965	1 228 500
Gross Profit	84 475	98 645	106 275	89 380	88 290	109 000	138 430	132 980	174 400	203 285	179 305	67 035	1 471 500
Operating Expenses													
Ashington													
Distribution	6 370	6 958	8 279	7 094	7 266	8 631	7 130	7 712	7 290	7 290	9 100	7 134	90 254
Factory expenses	29 452	30 057	32 738	29 199	27 643	31 411	22 033	30 627	27 380	29 095	35 372	27 248	352 255
Total Ashington	35 822	37 015	41 017	36 293	34 909	40 042	29 163	38 339	34 670	36 385	44 472	34 382	442 509
Eastbourne													
Distribution	327	327	408	327	327	408	327	327	408	327	327	410	4 250
Factory expenses	9 406	9 805	10 585	10 089	9 881	10 902	9 892	10 503	10 553	10 737	11 529	11 910	125 792
Total Eastbourne	9 733	10 132	10 993	10 416	10 208	11 310	10 219	10 830	10 961	11 064	11 856	12 320	130 042

Appendix VI   Budget Analysis 1979

Budget: 1979   Account No.::   Description: Summary 13 November 1978   **B**

	Jan	Feb	Mar	Apr	May	Jun	Jul	Aug	Sep	Oct	Nov	Dec	Total
Hatfield													
Research & Development	1 026	1 456	1 456	1 456	1 456	1 456	1 490	1 490	1 490	1 490	1 490	1 700	17 456
Sales expenses	14 999	14 472	14 742	14 675	14 162	14 257	14 757	14 745	15 175	15 448	15 280	14 951	177 663
Export expenses	4 848	8 571	3 848	4 620	4 746	5 635	5 752	3 886	6 963	3 880	4 264	5 897	62 910
Product development	2 701	1 921	4 861	1 771	2 461	2 741	12 234	5 874	3 124	1 594	2 754	2 599	44 635
Administration	10 317	9 770	9 070	9 083	8 770	8 980	10 044	9 632	9 282	9 015	8 952	10 699	113 614
Office and accounting	7 063	6 883	7 233	6 853	8 398	6 783	6 951	6 951	7 301	9 101	7 936	7 942	89 395
Head Off. Fac. Services	5 509	5 526	5 568	5 741	5 460	5 986	5 422	5 522	5 465	6 059	5 886	7 787	69 931
Total Hatfield	46 463	48 599	46 778	44 199	45 453	45 838	56 650	48 100	48 800	46 587	46 562	51 575	575 604
Total Operating Expenses	92 018	95 746	98 788	90 908	90 570	97 190	96 032	97 269	94 431	94 036	102 890	98 277	1 148 155
Operating Profit	(7 543)	2 899	7 487	(1 528)	(2 280)	11 810	42 398	35 711	79 969	109 249	76 415	(31 242)	323 345
Other (Income)/Expenses	12 219	10 940	11 508	12 043	14 307	16 260	16 079	15 928	16 372	15 023	14 464	12 102	167 245
Net Profit before Development	(19 762)	(8 041)	(4 021)	(13 571)	(16 587)	(4 450)	26 319	19 783	63 597	94 226	61 951	(43 344)	156 100
Development Costs	900	850	2 750	1 100	1 600	200	1 700	700	1 700	1 200	2 200	1 200	16 100
Net Profit before Taxation	(20 662)	(8 891)	(6 771)	(14 671)	(18 187)	(4 650)	24 619	19 083	61 897	93 026	59 751	(44 544)	140 000
Cumulative	(20 662)	(29 553)	(36 324)	(50 995)	(69 182)	(73 832)	(49 213)	(30 130)	31 767	124 793	184 544	140 000	

CHAPTER 4

The Bifurcated Engineering Group (Case History No. 2)

Since analysis and classification are essential tools of management research, management colleges may lead a student to suppose that there is but one main route to business success. Indeed, were human nature less variable this could, in part, be true. Fortunately, we escape the threat of bureaucratic monotony by the rich variation in our resources, environment, and nature. The interest of this particular case history, then, lies in its lack of conventional patterns and the resourcefulness of those who direct the company. It is an optimistic story, and the *dramatis personae* were not deterred by the economic gloom and forebodings that prevailed during the course of this study.

4.1   Introduction

Bifurcated Engineering Limited was set up in May 1969 as the holding company of the newly formed Bifurcated Engineering Group. The group today consists of seventeen subsidiary companies, all very successful and well established, engaged principally in the manufacture of rivets and other cold-forged products (see Figure 4.1), as well as rivet-setting machines, parts-feeding, and orienting equipment. The composition of the holding company board is shown in Appendix I, and short biographies of the chairman and group managing director are given in Appendix II.

The group's activities are centred mainly in the south-east of England and in the Midlands. The headquarters, and three of the subsidiaries, are at Aylesbury in Buckinghamshire. There is also a branch in South Africa and another in West Germany. (See Appendix III for details.) Within the past few months, a wholly owned subsidiary has been started in France.

4.2   The Aylesbury Factories

The principal subsidiary company in the group is The Bifurcated and Tubular Rivet Co. Limited for which the organization chart is shown in Appendix IV. This company was founded in Warrington, Lancashire, in 1892, and made bifurcated rivets for a wide variety of trades, including leather goods. With the upsurge of the then infant motor-car industries and consumer goods came an increased demand for their products, and in 1910 the company moved to its present site in Aylesbury, where it expanded both the manufacture of bifurcated and tubular rivets. The grandfather of the present chairman of the

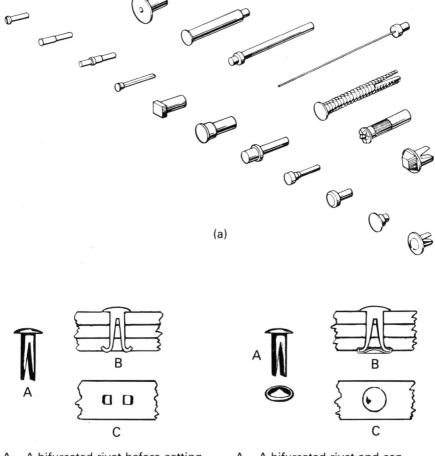

(a)

A. A bifurcated rivet before setting
B. Cross-section of material and
   clinched rivet
C. Plan view of the clinch

A. A bifurcated rivet and cap
   before setting
B. Cross-section of material and
   rivet with the prongs clinched
   in a cap
C. Plan view of the cap in position

(b)

Figure 4.1 (a) Special cold-forged parts; (b) The bifurcated rivet.
The bifurcated rivet is used as a self-piercing rivet in such materials as leather, plywood, vulcanized fibre, and plastics, and has the great advantage that, when set, the legs can be forced back into the surface of the material, leaving a relatively smooth surface. When fibreboard is riveted with bifurcated rivets the resulting joins are very much stronger than stitched joins, as the material itself has not been weakened by the large number of needle holes. These rivets can be used in automatic rivet-setting machines either by themselves or with caps, as shown

group obtained a controlling interest in the company at the turn of the century; the family atmosphere radiating from this holding has prevailed to this day, and is reflected in the excellent labour relations enjoyed throughout the group.

The centre of operations at Aylesbury is on an impressive 30-acre complex, containing the headquarters, the main subsidiary company, and two others. The site also includes its own sports club and playing-fields for a wide variety of games and activities. Today 'Bif & Tub' has the biggest production facility in the world, making semi-tubular and bifurcated rivets and other cold-forgings under one roof. It has a production capacity of around 20 million parts per day, and exports to most countries in the world. The managing director is responsible for all the companies on the Aylesbury site and reports directly to the group.

The cold-forged products of B. & T.R. are made on a large range of machines, most of which the company has designed and built themselves. Much of this machinery is very specialized, having many years of empirical experience built into it, and this process continues today, ensuring that the company is ahead technologically. Cold-forging enables the products to be manufactured to the consistently close tolerances necessary for mass-production assembly. In-house facilities also exist for heat-treatment, electro-plating, and other secondary operations.

The organization of the main factory at Aylesbury is extremely well laid-out; material-handling is minimized by careful arrangement of the departments. In the material stores, thousands of coils of wire of different materials—steel, aluminium, copper, brass, etc.—all lie in neat herringbone stacks, and this area is adjacent to the various production departments. After manufacture, the rivets or cold-forgings are heat-treated, plated in a wide variety of finishes, or painted as necessary, then packed ready for delivery. Inspection is on a continuous basis—from the raw materials, through all stages of manufacture, until final despatch. The buildings are tall and well-lit, gangways are clear and clean, and there is a general air of busy efficiency about the place.

In addition to rivets, B. & T.R. produce a range of riveting systems and setting machines. These vary from large gantry-supported machines having a large plate-width capacity, to smaller, hand- or pedal-operated machines for short production runs. For mass-production work automatic machines are available which feed, set, and complete the riveting operation.

Also at Aylesbury are two other companies in the B.E. group, the first of which is Aylesbury Turned Parts. This company was originally formed in 1925 to manufacture components for the aeronautical industry; it joined the B.E. group in the autumn of 1974. Aylesbury Turned Parts now produce a wide variety of precision-turned components and fasteners, in all materials, for the automotive and electronic industries, and provide manufacturing capacity beyond the cold-forging abilities of the other members of the group.

Figure 4.2   The Aylesbury centrifeed

In an adjacent factory is Aylesbury Automation Limited. This company specializes in automatic parts-feeding, orienting, and assembly machines, as illustrated in Figure 4.2. Started in 1961, it designs, develops, and manufactures a wide variety of feeders, assembly machines, and counting systems, level controls and conveyors for many industries, including food, electronics, and pharmaceuticals, and is an acknowledged leader in its field.

4.3 The Midland Group

Seven of the seventeen subsidiaries are situated in the Birmingham area, and for this reason a special Midland sub-group was set up in 1973 to develop greater rationalization and a closer working relationship between the companies located there.

Three of the Midland factories are also principally engaged in the manufacture of rivets and cold-forgings. The numbers employed vary from 50 to 100 people. Their turnover figures range from £400 000 to £1 200 000, and their net profits from 10 per cent to 30 per cent. They are essentially jobbing firms, making specialist parts in large and small production runs.

The largest is Black & Luff Limited, which, incidentally, was the first to amalgamate with B. & T.R. in 1960 and the managing director of that company, G.O. Luff, is today the chief executive director of the group. The company still manufactures semi-tubular and cold rivets, clevis pins, and precision cold-forgings. Next in size in the Midland group is Jesse Haywood & Co. Limited, who also make clevis pins, together with shoulder rivets. The third member, Clevedon Rivets & Tools Limited, specialize in non-ferrous rivets made from silver, titanium, and aluminium alloys.

Also in the Midlands area are three subsidiary companies engaged in activities peripheral to rivet manufacture. One of these is Holdfast Fixings Limited, who make special fixings for low-density materials (see Figure 4.3) and other applications in the building industry.

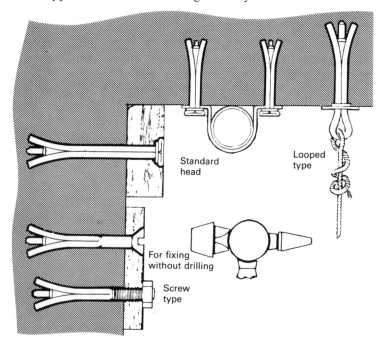

Figure 4.3   The Holdfast Loden Anchor

Another company, Weighpack Limited, joined the B.E. Group in 1975. At their Redditch factory Weighpack produce automatic weighing and counting systems for such industries as fastenings, confectionery, pharmaceuticals, and chemicals. Exports account for a significant part of the company's production, with well-established markets particularly in Europe, which are looked after by their associate company located in Holland.

The remaining Midland factory is GTN Printed Circuits Limited. A growing amount of B.E.'s products are for the electronics industry, from tiny terminal posts in their millions, to the feeding and orientating equipment necessary for their assembly, and this has recently been extended into the batch production of printed circuit boards. In-house facilities include design and drawing services, component assembly, and gold-plating.

4.4 The Remaining Subsidiaries

There are two other companies in the UK. One is Alpine Fasteners Limited, based at Letchworth, Hertfordshire, who are stockists and distributors of all kinds of industrial fasteners, including nuts and screws, and adhesives and sealants, although these products are not manufactured within the group. At Guildford in Surrey is another subsidiary, Dico Packaging Engineers Limited. This company specializes in the production of automatic closing and capping machines for all types of container, including the aerosol industry. It consistently exports approaching 35 per cent of its output, in the face of fierce foreign competition.

Bifurcated Engineering Limited has two main overseas subsidiaries, one in South Africa and one in Germany. The first of these, Sarmco, joined the group in 1973 and is based in Durban. There they manufacture rivets and also distribute most of the machinery manufactured by the UK subsidiaries. Sarmco also has a sub-branch in Johannesburg.

The group has had a long association with the Acme Rivet & Machine Corporation in New Jersey, USA, who have for many years attended to the affairs of the group throughout the United States. Acme Rivet & Machine Corporation are, in turn, a subsidiary of Elco Industries Inc. of Rockford, Illinois, and there is a constant interchange of technical information between the B.E. Group and these American associates.

Finally, in Hamburg, is a marketing and servicing organization for many of the group products in northern Europe. From its new office and warehouse, this company offers an on-the-spot service in one of the group's largest and most important export markets. The Hamburg premises also include an exhibition area.

4.5 Group Strategy

There are certain features of B.E.'s industry which make it very difficult to have a corporate strategic plan. First, it is a reactive industry—as opposed to being creative—and this usually makes it impossible to forecast which type of fastenings people might require in advance, so that little product planning is feasible. For example, a washing-machine manufacturer bringing out a new model would see no problems in joining together sections of the carcass— maybe they will be riveted together, maybe not, but not until the last stages of production of the machine would the rivet-manufacturer be likely to become involved, and then he must react quickly. Some years ago B.E. co-operated with a Belgian company to investigate the problem of predicting the demand for cold-forged products for the Continent. They carried out a joint market research survey which lasted for over two years, and the team included an economist and a marketing expert. Indeed, the survey was deep and very thorough. But the outcome of the investigation was a negative one; it was virtually impossible to forecast the demand for such fastenings—cold-forged or otherwise.

A second difficulty in forward planning stems from the fact that a considerable portion of B.E.'s output of rivets, etc. is required in relatively small quantities—but in a large range of shapes and sizes. Generally using customer-part numbers, B.E. manufacture many thousands of different types of rivet, a majority of which are in small runs. Clearly, therefore, it is not feasible to carry this great range in stock, and production schedules have to be planned and changed at relatively short notice. Flexibility, therefore, is the keynote in all its factories manufacturing cold-forged products.

Third, there is the nature of the business itself be considered. Although many of the rivet-setting machines designed and manufactured by B.E. are highly sophisticated, rivet-making in itself is not a high-technology industry. This is evidenced by the fact that the group suffers keen competition from numerous manufacturers, many operating with low overheads, who can pick and choose the most suitable items for their plant, and so offer a constant threat to the whole price structure of the industry.

To counteract these problems B.E.'s strategy is based on providing a complete and reliable service to all its customers, not only in the field of fasteners, machines, and assembly systems but also in the personal advisory service they provide to solve individual fastening problems. Experience and knowledge of material-flow properties is a vital part of the technology of cold-forging. Many of their customers are ill-informed on this subject, and B.E. spend much time in educating them on more feasible designs. B.E.'s 'outdoor' staff are not merely salesmen; they are technical advisers who have served apprenticeships in the factory and who are keen to help customers to solve their individual problems.

4.6 Acquisition Policy

Another major element in B.E.'s corporate strategy includes the acquisition of other companies whose business is closely associated with their own. Each company has been acquired as a logical extension of existing business and there has been no deliberate attempt to diversify into new fields. Not all these companies were profitable on acquisition, and it reflects favourably on B.E.'s management that, by improving the plant and providing greater motivation for the staff, most loss-making companies have been made profitable.

There are several advantages to this policy of acquisition. Obviously, the association lessens competition from that quarter and usually helps prices to be stabilized without seriously reflecting the mode of operations. The alliance generally strengthens the capabilities of all the companies by introducing new areas of expertise, plant, and capital resources. New market potential can often be realized, and the increased momentum, given the back-up of B.E., is often associated with a vigorous determination to develop and succeed on the part of management and the workforce.

Whilst now having seventeen subsidiaries the group is not a conglomerate but is rather like the cobbler who sticks to his own last, having expanded rapidly over the last decade through a series of acquisitions in its own or related fields. Virtually without exception, the initial approach has always been made by the companies wishing to be taken over, with the existing management remaining in the majority of instances.

Having made the acquisition of a new company the group makes every effort to obtain greater efficiency within that company, either by utilizing its output within the group or by ensuring that its range of products is brought to the attention of the many thousands of customers the group has overall. This is generally carried out by all sales and marketing departments in the various companies knowing the group's products and passing the relevant enquiries to the appropriate company. Rationalization of products, however, is not sought at the expense of major changes in the operation of the acquired company; in fact, it is group policy to operate the companies so that their original identity and character may be preserved. This attitude eliminates the need for many modern management techniques, such as job descriptions, formal staff assessments, technical forecasting, etc. Non-productive to production staff ratio is, therefore, low and few bureaucratic procedures are apparent at B.E.

All the companies are given a considerable degree of autonomy to operate under their respective general managers. Each is also free to exercise its own methods of working on the shop floor. Aylesbury, for instance, is orientated towards automation, whereas the Midland factories, who are more concerned with small engineering pieces, favour operator-aids. This system works satisfactorily, and, as long as each company continues to be profitable, headquarters does not interfere.

It should not, however, be construed that B.E. is complacent about improving its production efficiency. This is under constant review at all sections of the organization, and is particularly aimed at automating many of the labour-intensive processes. The use of microprocessors is now being considered for many of their machines; it is along these lines that future investment policy is planned.

4.7 Finance

Bifurcated Engineering Limited is a public company. The turnover of the group in 1978 was £13 206 658, and the net profit, before tax, was £1 589 277 (12 per cent of turnover). Approximately 18 per cent of the total turnover came directly from exports, which was equivalent to £2028 for each employee. During each of the last ten years B.E.'s turnover has consistently increased, and the overall profit growth has also been good, although there were setbacks in 1972 and 1975.

In 1978 the board paid a total dividend of 3.1p per share, an increase of 10 per cent, but although the accounts showed an increase in profits, due to inflation the situation in real terms was about the same as in the previous year. In his annual statement, the chairman commented: 'It is a case of running to stay in the same place, and that takes a great deal of effort on the part of everyone.' During 1978 sales of cold-forgings were at a level which enabled the company to mitigate the increased cost of overheads, wages, and raw materials. Machinery output, although satisfactory, was restricted through difficulties in obtaining skilled personnel. The group was involved in a programme of major capital expenditure aimed at increased productivity, and this was proceeding according to plan. The success of B.E. and its subsidiaries during the past ten years can be seen from the summary record shown in Appendix V. (Extract from Annual Report and Accounts for 1978.)

Within the group each company has a large measure of freedom regarding expenditure and cash flow, and each has someone trained in general accountancy principles who is responsible for the cash position and profits. Each subsidiary is well aware of the necessity to be profitable and, within a set period, to generate cash. Profits are remitted in total to a central fund, and each subsidiary receives a statement of profit-expectation from B.E. If one company's profits fail to come up to expectations, a holding company accountant (trained in a number of disciplines), carries out trouble-shooting operations. At B.E., Cost and Management Acountants are considered the backbone of financial control, and are preferred to Chartered Accountants. The former were trained on the shopfloor and are familiar with union negotiations and the general administration side of the business, whereas the latter are 'considered basically historians analysing what has gone wrong in the past'.

Capital expenditure on major items is under the control of the chief executive, taking the cash-generation position of the group into account. The general policy is that borrowings of the group will never exceed a fixed percentage of shareholders' funds. B.E. have a formula by which each company takes the previous year's depreciation and adds to it 10 per cent of the previous year's profit, and that sum is the permitted expenditure during the year for capital, research, etc. There is, in addition, a contingency fund arranged each year for new ventures and unexpected major items of plant.

4.8   Marketing

B.E.'s marketing policy appears surprisingly weak for a business with a turnover of more than £13 million. Admittedly, they do some advertising, and all their subsidiaries have either a sales manager or general manager who takes on this duty in conjunction with the group publicity manager. Within the group, however, they rely on fewer than twelve sales representatives. The reason for the apparent lack of drive in this area is that new business stems mainly from the recommendations of satisfied customers, and as this, in fact, generates a high percentage of the potential market in this country, it is felt that there is little need for a formal sales promotion.

The absence of market push is most marked among members of the Midland group. Here, there seems to be a very close match between existing market needs and company resources. Their policy of selling reliability and quality at a competitive price has proved so successful that there is little need to make strenuous efforts to seek new outlets. However, this apparent weakness has been recognized, and a senior marketing director has recently been appointed with a full brief to rectify this shortcoming on a group basis.

The relative smallness of the subsidiaries and the autonomy which they enjoy is often reflected in the enthusiasm of the employees regarding the generation of new business. The group managing director, Graham Luff, made the point that 'Nearly everyone in the company acted as a salesman every time he went shopping, when he looked for new applications for the products he made'. As an example, one employee noticed a fastening on his child's toy which would have been much improved by a riveted joint instead of the joint being used. He told his manager, the appropriate contacts were made, and a new order was established. This thinking also applies to the specialized machinery companies, where both the employees and the management are constantly on the lookout for products and outlets for the rapidly advancing high-technology automated plant which they now manufacture.

There are very many instances of business stemming from the loyalty and personal involvement of employees which would be unlikely to occur in a more bureaucratically controlled organization. For example, during a visit to the Clevedon Rivets & Tools Limited the general manager received us on time at 9 a.m., without comment, but we were later told by one of his staff that he had travelled to and returned from Somerset overnight in order to deliver an urgent consignment. Significant was the comment by Mr Warner of Black & Luff: 'Ninety per cent of our orders are delivered on time, and for the remaining 10 per cent of customers, notification is given well in advance of late deliveries.'

Approximately 20 per cent of group turnover comes directly from exports each year, of which about half go to Europe. From Aylesbury, no less than 500 million pieces are shipped every year to the USA. Dico averages 35 per cent of exports to overseas companies, nearly all consisting of bottle-capping machines. Aylesbury Automation export about 20 per cent of their output. On the subject of exports, Graham Luff explained that during overseas visits 'the export executive plays a most important function, especially in relation to export of technology'. One has recently returned from visits to East Europe, including Russia and Poland. The group manufactures in Australia and India, has a technical interchange with associates in the USA and West Germany, and has commenced manufacturing in Malawi.

4.9   Internal Relations

Bifurcated Engineering became a public company in 1948 but still retains the ethos of a family business, maintaining a friendly atmosphere throughout. The treatment of staff is very good and morale is consequently high. Amongst the benefits which the group offers is equal pension rights to any employee, whether staff or works, based on 1/60th of final salary for each year of service. It also gives long service leave of one extra month for every ten years of service. This extra leave can be accumulated, but 50 per cent of it must be taken prior to retirement; the balance may go towards an earlier retirement.

The company provides a very good sick-benefit scheme, and has an organization called the Silver Rivet Club for anyone who has served more than twenty-five years. After this length of service, all employees receive a Christmas hamper, together with a turkey. Silver Rivet Club old-age pensioners of the company continue to receive these benefits until death.

The company has its own apprentice-training scheme and ensures by continuous assessment that there is a path to the top. It has a number of directors and general managers who have taken this route.

Walking through one of the Aylesbury factories it was clear that there was a very friendly relationship between the group managing director and the staff, and he spoke to numerous people on the shopfloor by name, enquiring after their health and general activities, both inside and outside the company.

There is a marked loyalty, too, among members of the Midland group, and, although each firm is jealous of its own identity and reputation, they are all willing to uphold the efficient rationalization policy by passing on work to those in the group who have a comparative advantage in producing that particular speciality. The Midland companies hold fortnightly meetings of their executives: these are extremely valuable and generate a free flow of information on all aspects of their operations.

Between the Midlands group and headquarters, however, communications seem less good, and this is attributed to the fact that the two areas are sufficiently far away from each other to discourage frequent visits. Contact in general is limited to ten or so telephone calls per day between Birmingham and Aylesbury. In particular, the Birmingham companies do not appear to share the same perception of the corporate services available from Aylesbury as regards accounts, publicity, marketing, and engineering facilities. However, Midland pride may well play some part in this reluctance to make better use of management services. This problem of communication is not uncommon in group organizations; it is a question of achieving the right balance between interest and interference. Nevertheless, it cannot be easy to make the subsidiaries solely responsible for making a profit and, at the same time, impose a central policy.

The responsibility placed on the subsidiaries to be self-governing also creates some stress in the implementation of government legislation. In common with many other UK industries, B.E. have not been without their share of problems stemming from trade union pressures and government legislation on such matters as safety, noise, maternity allowances, redundancy payments, etc. Now there is a genuine concern about taking on more labour lest they should have to reduce it in future. By giving strenuous attention to efficiency, it has been possible to reduce the numbers employed at The Bifurcated & Tubular Rivet Co. Limited from 880 to 600 between 1972 and 1975.

The problem of labour relations is most marked in one of the Birmingham factories, where the workforce has a high percentage of West Indians and Asians, and language is a problem. Headquarters are aware of the problems, and naturally are anxious to mitigate them; to this end they have recently appointed a personnal officer to serve the Midland group.

4.10 Innovation

Bifurcated Engineering is not an innovative group in the conventional sense, i.e. it has no research department. It progresses through the evolution of traditional products and the ability to recognize ,and absorb new businesses—a policy which is strongly supported by the shrewd judgement and entrepreneurism of the group managing director. Two examples of this 'business innovation' policy are worth noting as they typify this opportunist policy. Some years ago one of B.E.'s more enterprising employees, whose hobby was electronics, developed a particular interest in the design and manufacture of printed circuit boards. So intense was the enthusiasm of that particular individual for his hobby, that it eventually attracted the attention of his employers through the adverse effect it was having on his work. The attitude of most companies to this situation would possibly have been one of disapproval. The management, however, were aware that no business equals the success rate of an entrepreneur's own activities and took the opposite view; they encouraged the expansion of this hobby into a business and even provided facilities for its development. In addition, they made an agreement with the employee to buy out his business when he had reached an annual turnover of £100 000. This did not take long to achieve, and today GTN Printed Circuits Limited is operating as a subsidiary of B.E., from its own factory in Sutton Coldfield.

A similar situation arose out of a chance meeting in 1971 between the group managing director and an inventor by the name of George Webster. Mr

Webster was a modest man, and was reluctant to call himself an engineer; he had at various times farmed, run a holiday camp, and engaged in a variety of other miscellaneous jobs. In his early fifties, Mr Webster was inventive to a high degree, and said his ideas came while working or engaged on matters other than work. Some time earlier, his attention had been drawn to the need for a method of fixing into low-density materials, such as Thermalite building blocks, etc., and he had devised an ingenious fastener for this purpose which simply expanded inside the material when it was hammered in. He had three other devices which he had developed to the prototype stage, and he had gone to Aylesbury to look for premises in which to begin manufacturing. He had two other partners with whom he had started a small company.

In Aylesbury he met B.E.'s group managing director, who immediately recognized the potential of these devices for the building trade, and was so intrigued by them that he offered to take Mr Webster and his small company into the B.E. Group. In 1974 an agreement was signed with a company (internationally known in the DIY field) giving them a franchise for marketing. The contract was severed in 1977, however, because the stockists of the company had so many products that B.E.'s Loden Anchor was given insufficient publicity. Today the company operates as Holdfast Fixings Limited, in Redditch.

These two examples illustrate one of the major strengths of B.E.'s development policy; active product evolution. The ingredients for success in this direction are quite different from those of conventional attempts at innovation. B.E. has no chartered engineers on its staff (although admitting to the possibility of being disadvantaged by this); they have succeeded by virtue of being level-headed shrewd businessmen, whose first priority is to respond to customers' demands with regard to service, quality, and price—in that order. They obtained the right machines for achieving these ideas—and they worked them hard.

Although, as stated, B.E. do not have a research department they do have a very lively development area at Aylesbury. Work in this department is mainly concerned with the development of new machines for counting, orientating, and assembly processes. Skilled mechanical and electronic engineers are working on the development of prototype units—many being constructed in modular form. Eventually these systems will be incorporated into the production lines of the relevant subsidiaries. More recently, the policy has been to use resources outside the group, and development contracts have been placed with a polytechnic and a university.

As well as the new development of assembly systems B.E. are continually faced with new problems associated with their cold-forging processes. Components required not merely for fastening but as stepped pivots, control guides, etc. all need sophisticated tooling and machine systems to provide the sequential forming operations necessary for economical production. Again, a separate part of the factory is set aside for the development of prototype equipment.

These, then, are the ways in which B.E. is innovative—by reacting to the needs of their customers and initiating new projects and new products for which there is a specific order, as opposed to those for which new markets will have to be found. This policy of evolutionary development by responding to user-need clearly is more economically stable than a policy of innovation which might involve moving into new venture areas, with the high risk and financial outlay such ventures usually involve.

4.11 Conclusions

There are several reasons for B.E.'s success as a group. Their Aylesbury factories are very well planned and organized and, although more distant from the Midlands group than management might prefer, the main body of the organization, including headquarters, is contained on the one campus and the direction of all group activities from that centre works very well.

B.E. are fortunate, too, in the make-up of their board and management team. It is these men whose policies and decisions created the right atmosphere to produce that most vital ingredient for commercial success—a stable workforce; and it is their acumen which has enabled the firm to develop its business in a way which brings new orders, mainly from previously satisfied customers. The group managing director contended that 'There is no business in the world so good that it cannot be improved by better management'.

Management apart, however, there are two factors inherent in B.E.'s particular business which are fundamental to its continuing success. First, the market; there are very few industries today which do not require fastenings, and riveting has much to commend it. It is simple, effective, clean, and cheap. It requires no special skills, and where machines are needed to facilitate the process, these are readily available—from B.E. of course. Rivets can serve as fasteners, pivots, terminals, etc. The process can be used to fix any two dissimilar materials, from leather to plastics. Few of these claims can be made for either welding or adhesives.

Second, there are inherent economic advantages in cold-forgings. The process is economical in that there is virtually no waste—100 kg of wire will produce 100 kg of rivets. Production is automatic and extremely rapid, yielding products of good surface finish and close tolerances. Cold-forged parts have greater strength than comparable turned or moulded parts by virtue of the beneficial gain-flow when head-forming, etc. These two fundamental advantages—market potential and process efficiency—clearly provide B.E. with the right foundations on which to build a successful business.

But what of their weaknesses and threats, and how have B.E. so far managed to stave off these?

First, consider their competitors. Cold-forging of small simple items like rivets is not a highly specialized operation, so, not surprisingly perhaps, a large part of B.E.'s competition comes from numerous small companies who can operate on very low overheads. By selecting particular items, the small companies can produce these in large quantities and so threaten the pricing structures of the industry. There are two other major sources of competition in the UK, both much smaller than B.E. but both of whom make and sell rivet-setting machines. In Europe, there are no known companies who offer this complete service, as do B.E., so competition from abroad is minimal. There were some imports of rivets although these were not regarded seriously due to the high weight-to-cost ratio for most parts. Nevertheless, a company which can export 20 per cent of its production must be vulnerable to some degree to the possibility of imports.

Clearly, B.E. by no means hold a monopoly on the rivet-making industry, and cannot afford to be complacent about their competitors if they wish to retain their lead in this well-established field.

Finally, a possible threat may rise from revolutionary improvements in bonding techniques. With the advances currently being made in adhesives, and the possible spin-off from molecular research, such a possibility should not be overlooked. Asked whether they had any contingency plans for such an event, Graham Luff said they had not. 'But', he added, 'I generally find when one door closes, another opens. Whilst I am an eternal optimist, this is an area on which we keep a very close watch'. It is this kind of philosophy which is inbred into the company's policy, a philosophy which has served them well for more than three quarters of a century, and will almost certainly continue to do so.

Appendix 1   Holding Company Board

J.M.A. Paterson, JP, MA	Chairman*
G.O Luff	Group managing director
D. Whitehead	Company secretary†
G. Kershaw	Non-executive* (MD of The Bifurcated & Tubular Rivet Co. Ltd)
R.I. Paterson, MA	Non-executive*
R. Gabriel, MA	Non-executive*

*An engineer and trained in management
†Financial director and company secretary

Appendix II   Biographies of the Chairman and Group Managing Director

PATERSON, John Mower Alexander, MA (Cantab).
Chairman Bifurcated Engineering since May 1969.

B. 9 November 1920. *Educ.* Oundle School, Cambridge University, Honours degree, Mechanical Sciences Tripos, MA.

Career Details: Army, 1941-6; Director and works manager, Bifurcated Tubular Rivet, 1948-60; Chairman and managing director, 1960-9; 1974 to date, Chairman, Bifurcated Engineering Limited.
Clubs: Royal Ocean Racing, Royal Lymington Yacht.

LUFF, Graham Oliver.
Managing Director, Bifurcated Engineering, since May, 1973.
B. 7 September 1923. *Educ.* Bishop Gore's Grammar School, Swansea.

Career Details: Joint-Founder Black & Luff, 1948; Works director, 1952-64; Managing director, 1964-9, Managing director, The Bifurcated & Tubular Rivet Co. 1969-73.

Appendix III The Bifurcated Engineering Group

Holding Company

Bifurcated Engineering Ltd	Aylesbury, Bucks

Subsidiary Companies

Alpine Fasteners Ltd	Letchworth, Herts
Aylesbury Automation Ltd	Aylesbury, Bucks
Aylesbury Turned Parts	Aylesbury, Bucks
The Bifurcated & Tubular Rivet Co. Ltd	Aylesbury, Bucks
Bifurcated Engineering (France) SA	Noisy-Le-Grand, France
Bifurcated Engineering GmbH	Hamburg, Germany
Black & Luff Ltd	Kings Norton, Birmingham
Clevedon Rivets & Tools Ltd	Kings Norton, Birmingham
Craig Engineering Ltd	Vale, Guernsey
Craig Marketing Ltd	Fareham, Hants
Dico Packaging Engineers Ltd	Guildford, Surrey
GTN Printed Circuits Ltd	Sutton Coldfield, West Midlands
Jesse Haywood & Co. Ltd	Smethwick, Warley, West Midlands
Holdfast Fixings Ltd	Redditch, Worcs
Pearson & Beck Ltd	Redditch, Worcs
Sarmco Africa (Pty.) Ltd	Durban, South Africa
Weighpack Ltd	Redditch, Worcs

Appendix IV Bifurcated Engineering Limited Board

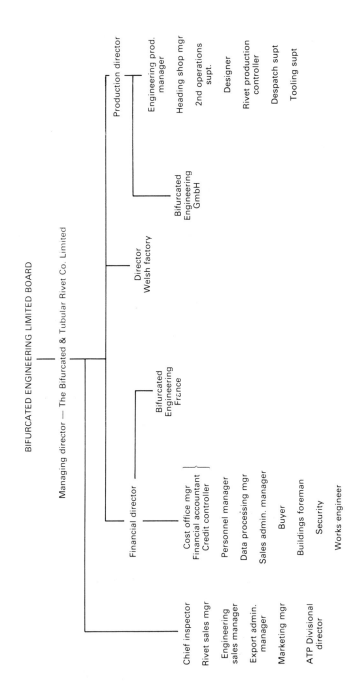

BIFURCATED ENGINEERING LIMITED BOARD

Managing director — The Bifurcated & Tubular Rivet Co. Limited

Financial director

Cost office mgr
Financial accountant
Credit controller

Personnel manager

Data processing mgr

Sales admin. manager

Buyer

Buildings foreman

Security

Works engineer

Chief inspector

Rivet sales mgr

Engineering sales manager

Export admin. manager

Marketing mgr

ATP Divisional director

Bifurcated Engineering France

Director Welsh factory

Bifurcated Engineering GmbH

Production director

Engineering prod. manager

Heading shop mgr

2nd operations supt.

Designer

Rivet production controller

Despatch supt

Tooling supt

Appendix V Bifurcated Engineering Limited and Subsidiaries

Ten-Year Record

£'000	1978	1977	1976	1975	1974	1973	1972	1971	1970	1969
Turnover	13 207	12 202	9841	8223	8131	6420	5010	4489	4434	3939
Trading profit	2056	1934	1740	1307	1431	1123	782	901	816	800
Depreciation	407	373	356	355	345	338	273	257	227	195
Interest payable	60	85	127	184	162	70	40	44	25	16
Exceptional items		80	115	130	60	45	48	15	41	
Profit before taxation	1589	1396	1142	638	864	670	421	585	523	589
Taxation	812	725	514	345	434	306	165	197	209	256
Earnings	777	671	628	293	430	364	256	388	314	333
Extraordinary items		81		10	(8)	59				
Dividends net	233	212	191	172	161	157	146	129	108	105
Income tax on dividends							30	81	75	73
Earnings retained	544	378	437	111	277	148	80	178	131	155
Assets employed										
Fixed assets	2997	2728	2678	2855	2839	2658	2223	2278	2146	1991
Trade investments	7	6	32	43	44	44	54	56	65	13
Goodwill	389	387	424	420	437	413	277	272	88	28
Net current assets	3694	3265	2507	1593	1244	817	1004	902	931	1031
Total assets employed	7087	6386	5641	4911	4564	3932	3558	3508	3230	3063
Financed by										
Share capital	1881	1881	1567	1567	1567	1567	1567	1567	1442	1406
Group reserves	3454	2898	2871	2372	2162	1877	1718	1625	1256	1093
Minority interests						4				
Investment grants and deferred taxation	1752	1607	1203	972	835	484	273	316	532	564
Total funds invested	7087	6386	5641	4911	4564	3932	3558	3508	3230	3063
Earnings per share	10.33p	8.92p	8.35p	3.90p	5.72p	4.84p	3.41p	5.16p	4.54p	4.93p
Gross dividend per share	4.624p	4.224p	3.878p	3.526p	3.264p	3.063p	2.92p	2.92p	2.92p	2.72p

CHAPTER 5

English Glass Company Limited (Case History No. 3)

Small companies not only display great diversity among themselves but also within their own activities. This case history describes a company which grew from two employees some fifty years ago to around 150 employees in 1980. The company showed considerable acumen in dealing with a range of situations which are popularly believed to be the prerogative of large organizations.

The original product was launched to challenge the complete domination of a market segment by an imported product, and during the next six years a range of products were sold based on evolutionary development. The wheel of fortune then turned a complete circle and the company was faced with a flood of cheap imports which threatened their main lines. Its response was to replace evolution with broadly based diversification and it quickly took out a patent on a new product that lay well outside its basic range. Shortly afterwards the company acquired a license for manufacturing a third type of product and obtained a marketing agency for a fourth. However, market conditions are rarely stable for long, and within a short span they had to sell manufacturing rights of their patented article. So here we have a small company who, having progressed from evolution to innovation, is now manufacturing a licensed product, operating a marketing agency, and has carried out a successful divestment.

5.1 Introduction

English Glass Company Limited (Englass) has been established for nearly half a century. 'We started with two men pressing glass reflectors from rod', said Tom Lawson, the managing director. 'This year we will sell a diverse range of products all over the world for £4m.' Englass has a range of industrial glass products (its original sphere), a wide variety of dispensers manufactured under license, and it markets technical ceramics. It supplies these products to a wide spectrum of UK and overseas customers. It employs some 140 people in premises with a total floor area of about 4500 m².

This case traces the development from a private limited company in 1934, employing two men, to an organization with a balance of manufacturing and factoring in three quite different product areas. The interest of the case centres on the retention of profitable small batch orders to meet specific customer-requirements and some rationalization of existing products, together with the development of new products and new markets. This involves balances between new in-house production and continuing the use of sub-contractors, having regard to the labour-intensiveness of the firm.

Where the balance lies is at the heart of management's forward planning and it determines the viability of automation in certain areas. It will be seen that, during the 1970s, Englass' policy was less affected by financial constraints than many other industrial firms but, to a firm with a high export ratio and a high labour factor, the current combination of high inflation and the strengthening of the pound presents new challenges for the 1980s. These factors have caused the board to put in train the preparation of a corporate plan for the future. In 1980 the directors, in conjunction with consultants, drew up a corporate strategy specifying company targets. This case reviews some of the options against the background of the Englass organization and resources.

5.2 Historical Background

English Glass was formed in 1934 by two directors of the John Bull Rubber Company to manufacture glass reflectors for motor cycles and cycle accessories. Previously there was no manufacturer of these products in the UK. Reliance had been on imports from Czechoslovakia. By 1940, when the trade name Englass was adopted, production had diversified to embrace glass beads, hat pins, stones for jewellery, switchboard lenses for the Post Office, and camera viewfinder lenses. During World War II the firm made millions of acid-filled glass ampoules for mines and boobytraps.

For the first ten years of the post-war period the fancy jewellery business continued to be profitable but, in the mid-1950s, the market became flooded with cheap foreign imports and fashions changed. In the late 1940s English Glass patented a metal 'Catseye' type of reflector for the centre of the road. This achieved modest sales at first. But sales to Africa increased for, unlike the conventional rubber-mounted 'Catseyes', the glass lenses were not easily extractable for making into necklaces, etc. The know-how in road-markers and glass was eventually used in the early 1960s to develop road-markers which could be stuck onto the road-surface instead of having to make a hole in the road. English Glass pioneered this stick-on system which is still in use throughout Britain today.

To counter the loss of the fancy jewellery market the firm developed during the 1950s the road-mouldings for products in the optical, electrical, medical, and general engineering fields. The manufactured-glass range has, over the last twenty-five years, been progressively extended. Examples of new activity are machining of glass components for a variety of applications and specialist lines for esoteric customer-requirements.

In 1960 English Glass moved to its present site at Scudamore Road, Leicester. Its premises have grown substantially over the twenty-year period. The floor area has increased from 2500 to 4500 m^2 in the last three years (see Figure 5.1). In the early 1960s the firm, faced with escalating costs and growing competition, actively sought diversification. It looked for new areas of business, new products, and new markets for existing products. The two

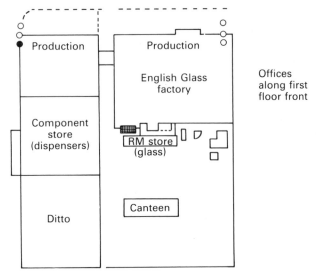

Figure 5.1   Factory plan

most important results from this were the acquisition in 1965 of the exclusive manufacturing and marketing licence from the Bakan Plastics Division of the Realax Corporation of Kansas City, Missouri, USA, for their range of spraying and dispensing pumps; and becoming, in the late 1960s, the exclusive UK marketing agency for the Wade Group range of high-alumina ceramics. This latter arrangement came about after the formation of an Englass ceramic division in 1964, born out of its expertise in the glass substrate field. After two years of development of glass-based products and market research it was concluded that greater potential lay in the high-alumina ceramic material and, hence, the marketing experience of Englass and the manufacturing capability of Wade came together.

The firm entered the 1970s with four divisions concerned with quite different products: industrial glass, road-markers, dispensers, and ceramics. The road-marker activity was sold off in 1970. This decision was taken because of the retirement of the then managing director, who had a specialized knowledge of this process, and also because of the high administrative burden to a small firm, selling this product in small quantities, to a potential market of 600 local authorities.

So during the 1970s there were three divisions: one manufacturing the firm's own developed products, one manufacturing under license, and one marketing. As the decade progressed, factoring and manufacturing cross-fertilized. Labour costs were continuously monitored, subcontracting was used, and expertise bought in to avoid wide fluctuations in the number of employees as market demand changed.

At the start of the 1980s the firm is, as it was at the beginning of the 1960s, faced with more financial constraints than in the previous decade and a similar need to rationalize and diversify. Taking opportunities in new fields worked then, and so the provisional target is that 80 per cent of the firm's profits in 1990 will be from products not currently manufactured or marketed. It is another provisional aim that the current export ratio of 40 per cent be increased to 70 per cent. The above targets, together with other relevant objectives, are currently under discussion by the directors as part of the corporate planning exercise.

5.3   Product Range

Industrial glass

Some of the basic items manufactured by Englass are precision-ground glass balls, glass rod mouldings, mirrors and lenses, and Ballotini (small glass balls). Adaptation of these items to applications to meet particular customer-demands means incorporating them into field strip viewers, airfield landing lights, foam injection systems for cavity-wall installations, optical systems, and beds for patients suffering from burns or bed-sores.

There is a flexibility to modify the composition of the product to meet a new customer-requirement. An example of this is the specification of ICI for a non-alkaline Ballotini product for grinding chemical compounds. The standard Ballotini has a specific gravity of 2.9 because the greater weight reduces grinding time substantially. This feature was suitable for ICI's purpose, but the standard Ballotini was not, as its alkaline content could

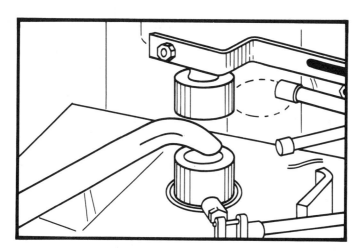

Figure 5.2   The craft of glass-rod moulding: a molten glass rod
is placed between two mould surfaces

82

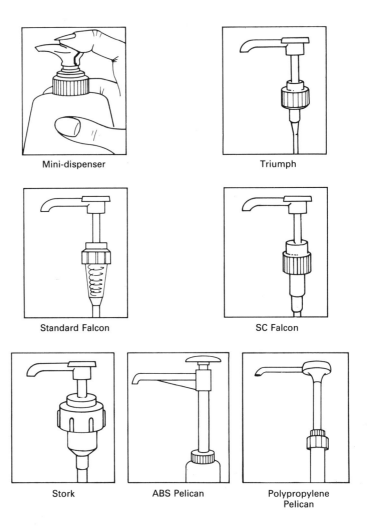

Figure 5.3   Dispenser pump range

cause damage to, or even decompose, compounds being milled. Non-alkaline Ballotini, similar to the standard in other respects, was supplied. One of the manufactured processes, shown in Figure 5.2, illustrates the craft origins of the industrial glass process. A comprehensive list of products using Englass manufactured industrial glass is given in Appendix 1.

Dispensers

Although there are perhaps sixty varieties of dispenser, with applications from polishes to pesticides and from mustard to mouthwash, they are effectively variations on seven themes, as shown in Figure 5.3. Again, there is

Figure 5.4   Snapsac

a wide variety of applications for dispensing of a standard dose, from catering uses—ketchup, mayonnaise, cooking oils—to hospital use as skin-cleansers for bathing babies, promoting safety, and economy. There are domestic horticultural applications for plant feeding and agricultural uses for weeding.

A product of the dispenser division not involving a pump is the Snapsac, shown in Figure 5.4. This is a wastebag holder for confined spaces, and can be used in homes, caravans, boats, hospitals, and laboratories. The snap closing action effectively seals in smells and so improves hygiene.

Ceramics

As has been stated, ceramic products marketed by Englass are manufactured by the George Wade Group, at the Burslem factory or the Wade (Northern Ireland) factory at Portadown. In all publicity the products are normally referred to as Wade/Englass products. A more comprehensive list of products using Wade/Englass items is given in Appendix II.

5.4   Organization

The organization is headed by the board of management, consisting of five members, and is supported by an advisory committee. The composition of the board is a non-executive part-time chairman and four full-time executive directors. The four full-time executive board members are the managing director, the sales director, the works director, and the company secretary. Biographies of the board members are given in Appendix III, and a breakdown of their areas of responsibility and the compositon of the supporting advisory committee is shown at Appendix IV(A). More detailed analyses of the organization under each of the four executive directors are given in Appendices Iv(B) to IV(E).

The organization has recently become more function-orientated than product-orientated, although there are senior managers who have responsibility for only part of the product range. The managing director has overall responsibility for all products and functions. The company secretary-cum-financial director is responsible for the financial side of the whole company.

A very recent rationalization move combined the glass and ceramics divisions into one division under the sales director. The general sales manager has the sales responibility for the dispenser division. The technical manager's job encompasses, primarily, new product development, quality control, and all technical services for the dispenser division. The material control and purchasing functions are the responsibility of the material controller and cover all glass and dispenser products, but not ceramic items, the purchase of which items is the responsibility of the sales director. Manufacturing at English Glass is the responsibility of the works director and covers glass and dispenser products.

A very important feature of the dispenser division's sales organization is the use of agents or distributors. There are twenty-four of these, covering the major selling areas in Europe spread geographically within a triangle approximately bounded by Helsinki, Lisbon, and Tel Aviv. There are also agents or distributors working for the glass and ceramics division in Germany, France, and Switzerland (all agents for ceramics) and in the USA (glass division distributors).

It will be seen from the Department of Employment comparison in Appendix V that there were 134 employees in a ratio of five females to four males in 1979. Forty-seven per cent were engaged directly in production, together with a further 11 per cent in supporting services to production. The balance of 42 per cent was engaged in non-technical functions, such as purchasing, sales, finance, and general administration.

There is a determined policy on age-spread among key staff. The purchasing officer is 34; the commercial manager is 35; the technical manager is 41; the works manager is 48; the management accountant is 49; the general sales manager is 51; and the cost accountant is 57.

5.5 Finance

Performance in 1979, distribution and profits, and comparative data over the period 1975 to 1979, are given in Figures 5.5–5.7. These are extracts from an information circular to all employees on how the company is progressing. Whereas sales volume was less than £3m in 1978, and approximately £3⅓ m in 1979, the forecasted sales for 1980 are over £4m. Some 50 per cent of dispensers, 25 per cent of ceramics, and 25 per cent of industrial glass are

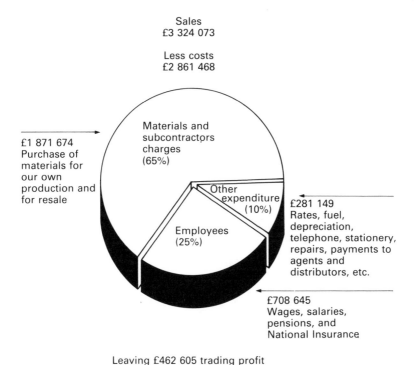

Sales
£3 324 073

Less costs
£2 861 468

£1 871 674
Purchase of
materials for
our own
production and
for resale

Materials and
subcontractors
charges
(65%)

Other
expenditure
(10%)

Employees
(25%)

£281 149
Rates, fuel,
depreciation,
telephone, stationery,
repairs, payments to
agents and
distributors, etc.

£708 645
Wages, salaries,
pensions, and
National Insurance

Leaving £462 605 trading profit

Figure 5.5   Sales and costs 1979

exported to twenty-nine countries spread over five continents. Total assets at 31 December 1979 were £1 109 000 financed by share capital of £178 000, various reserves £925 000, and secured loans of £6 000. The largest shareholder is the Industrial and Commercial Finance Corporation Ltd, a company formed in 1945 to provide a source of longer-term finance for small and medium-sized businesses and which is owned by the Bank of England, the English clearing banks, and the Scottish banks.

5.6   The Current Situation

As shown in Figure 5.7 and the forward estimates, there has been a significant growth of assets, sales, and profits over the last five years. Englass is, therefore, entering the 1980s with a sound financial base. The current economic climate is significantly less favourable to a small labour-intensive firm with high exports.

Looking at this in more depth, 55 per cent of dispenser exports, or 28 per cent of total output, goes to Western Europe. Perhaps half the European exports go to West Germany. Inflation in the UK is running at four times the

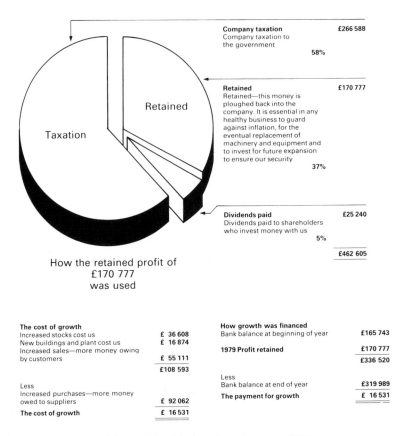

Company taxation Company taxation to the government	**£266 588**
58%	
Retained Retained—this money is ploughed back into the company. It is essential in any healthy business to guard against inflation, for the eventual replacement of machinery and equipment and to invest for future expansion to ensure our security	**£170 777**
37%	
Dividends paid Dividends paid to shareholders who invest money with us	**£25 240**
5%	
	£462 605

How the retained profit of
£170 777
was used

The cost of growth		How growth was financed	
Increased stocks cost us	£ 36 608	Bank balance at beginning of year	**£165 743**
New buildings and plant cost us	£ 16 874		
Increased sales—more money owing by custsomers	£ 55 111	**1979 Profit retained**	**£170 777**
	£108 593		£336 520
		Less	
Less		Bank balance at end of year	**£319 989**
Increased purchases—more money owed to suppliers	£ 92 062	**The payment for growth**	£ 16 531
The cost of growth	£ 16 531		

Figure 5.6   Distribution of profits 1979

West German level. To remain competitive in West Germany, the annual increase in charge in deutschmarks cannot exceed 5 per cent, but increased production costs are likely to approach 20 per cent. The effect of this on profits for a constant sale level is obvious. Over the 1974–8 period revenue from sales doubled but profits only went up from 10 per cent to 13 per cent. If labour costs increased by 20 per cent per annum and other costs (materials, fuel, buildings, rates, and office services) by 15 per cent, the 13 per cent profit would reduce to 3 per cent. To restore the balance, either sales need to be multiplied in the order of 8:1 or costs reduced by 16 per cent, or some equivalent combination of sales increase and cost decrease would need to be achieved to maintain present levels of profit rather than to continue annual growth.

It is imperative that this problem be satisfactorily solved, for profitability is dependent upon export business. There is a need to export to support the expansion of home markets. Although inflation and a strong pound makes exporting difficult, the problem would be accentuated if the UK left the EEC (which takes half these exports) and favourable duty rates were lost.

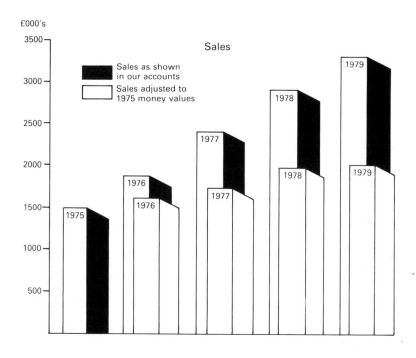

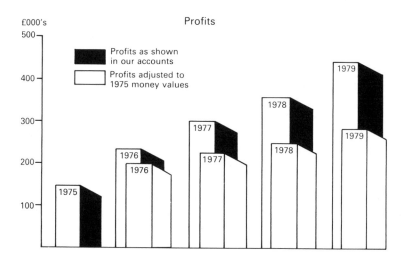

Figure 5.7   Sales and profits 1975–9

It is acknowledged that this analysis is an over-simplification but it does point to the order of the challenge to management. The lack of financial constraint during the prosperous 1970s has made rectifying action to retain profit levels more urgent today, for it has made the company unwilling to expand and encouraged it to reduce overheads stringently.

To a degree, the ability to restrict overheads is limited by the diversity of the business. This company, with less than 150 employees, has three divisions producing and/or selling quite different types of product. Moreover, each division has a different *modus operandi,* needing variations in strategy for its success. To some extent, however, the number of direct employees is not an exact criterion for the size of the firm. Its sales and marketing organization is backed by an extensive agent and distribution organization. There is no drawing office; normally one draughtsman and sketches from customers suffice. Design for the dispenser division's products is bought in from consultants. The external support for the non-technical areas such as sales, provided by the twenty-four overseas agents and distributors, involves significantly more people than the limited support for the technical side from design advisers.

It will be seen in Appendix V (column 30.12.79) that the sixty production employees are supported by three in engineering and fifteen in 'indirect works', which comprise, *inter alia,* work administration, warehouse, despatch, and raw materials. Of the remainder, five are technical, which means that some 38 per cent of the emloyees are not involved in production. This high ratio is much-influenced by the growth of factoring as compared with manufacturing. There is no manufacture of ceramics, and the factoring share of total business has remained at about 50 per cent.

While it is not suggested that staffing on the non-technical side is extravagant—sales staff have only increased by one third while turnover has doubled—the proportion of staff who are non-technical and not involved in production is high. A reduction of this proportion would reduce overheads and improve profitability. A potential means of doing this, without loss of efficiency or damage to staff relations, is provided by the installation of a computer. The role of the computer, and the prospect it offers of reducing non-technical labour costs, is described later in this section (page 90).

It was interesting to learn that whereas industrial glass was given a ten-year life in 1970, the sales, in fact, remained approximately constant. This is due to the fact that, although business at traditional outlets fell, the difference was just covered by the discovery of new applications. The drop in sales was attributable to foreign competition and a growing popularity of photographic prints; and the gains came from such items as condensers for microfilm readers and lenses for photoelectric applications. There are no indications that the turnover plateau can be indefinitely sustained. In contrast, there are some twenty new developments in hand on the dispenser side and some new contractual possibilities which show promise of successful entry into the injection-moulding field. Dispensers have a strong manufacturing base, for the percentage of manufacturing to factoring is 85:15. Although some UK manufacturers make products similar to one or other of the Englass pumps, there is no UK across-the-board competitor. Two other problems are that first, most manufacturing runs are in small batches and second, the bulk of business is now manufacturing under license or marketing the products of other firms.

A significant proportion of manufacture must meet special customer-requirements. It is essential to the survival of a small firm that it makes products which cannot be produced economically by the customer-firm or by its larger sub-contractors. If, however, it is not selling larger batches to regular customers or making standard lines there are obvious problems of overheads. Scope for automation or new plants is restricted. Without some rationalization of products, perhaps by a rise in the size of minimum order, reduced labour-intensiveness is difficult to accomplish. Stockpiling, too, without some larger or standard orders becomes too small an operation effectively to use labour during trough periods thereby introducing a nugatory overhead.

In principle, dependence on licensing and contracts carries with it some vulnerability. It is reasonable to accept the loss of such licenses and contracts if one fails to deliver the goods, but, if one is successful, one can take business to the point where a principal decides that it is more profitable to manufacture and/or market the product himself. For example, a contract to act as agent in the UK for a West German firm concerned with optical design moulding, grinding, and polishing glass was terminated after the business had achieved a contribution towards Englass industrial glass of slightly less than 10 per cent, because it was concluded that the business was now big enough to operate its own representative in the UK.

The main contracts with Wade and Realex are very much more secure than this. There are now Wade/Englass products on a closely integrated basis and there are obvious advantages to a US firm such as Realex in terms of differences in marketing techniques if it can use an effective licensee in the host country. In aggregate, this calls for a corporate plan and an associated strategy for the next ten years. The firm has this in train, with a more specific and detailed plan for the 1980–4 period. The management's broad policy on expansion is to improve productivity, complement or replace narrowly based products, improve sales outlets, and explore new licensing and agency opportunities. There is current action as well as forward planning in this context. The company has prepared new product-search specifications for each of their three project ranges and has held synectics sessions run by external specialists as a means of assessing new products and market potential.

Some new developments have been mentioned, and these may involve shift working, which has not applied hitherto, and may introduce recruitment and/or supervision problems. Additionally, marketing has been strengthened and an important step towards reduction in the non-technical labour costs will start in September 1980 with the installation of a computer.

Reference has been made to the European Agency and Distribution organization. Each agent or distributor visits the UK about once a year. The general sales manager visits each agent annually and spends seventy days a year exclusively on export business. Every three years there is a conference for agents and distributors at Düsseldorf during an international exhibition. This is now being strengthened by the conversion of agents who only solicit business to distributors who sell in their areas.

It has been mentioned that when there was a similar need to expand to meet escalating costs twenty years ago, new opportunities were seized to achieve the goals which bore fruit in the 1970s. Management recognizes that a repetition of that success cannot be guaranteed, so it has strengthened market research as distinct from sales by the appointment of a marketing officer (in the dispenser division) working closely with the technical manager on new projects. Beyond this, there is an interesting marketing experiment aimed at improving the effectiveness of agents and distributors. Englass has made use of a marketing course at the North Staffordshire Polytechnic for their language graduates. Two graduates were sent to the countries of their language for four weeks to provide an independent appraisal of the effectiveness of overseas selling agencies and distributors. Their lecturer visited them during the stay. Costs are being shared by Englass and its distributors and the polytechnic. The report subsequently received indicates that a fundamental change in method of sales in certain European countries will have to be undertaken.

The computer will be used for financial accounts, stock records, sales/order processes and invoicing, material control, and cost accounting. It offers the prospect of savings for non-technical staff, especially at the lower level, where there is a significant natural wastage. To improve efficiency, it could also reduce overheads in a less direct way. At present, cost accounting requires much interdepartmental consultation. To avoid complication, the same system for small and large batch orders is used. As a result the procedure is probably unnecessarily sophisticated for the small orders, where there is less competition, and not sophisticated enough for the larger batches, where competition is high. The computer should enable greater precision, with less staff effort, and improved competitiveness through less need, having obtained precision, to build in safety factors for contingencies.

These are some of the pieces in the jigsaw which is now the priority task of management to put together. It is now appropriate to look at the options open to Englass to make sure the growth of the 1970s is sustained in the less favourable climate which now prevails.

5.7 The Way Ahead—The Options

The (provisional) management target that 80 per cent of profit in 1990 should derive from new products and the policy to complement or replace narrowly based products are important pointers to the way ahead. Retention of profit levels similar to the last five years, let alone improved profitability, depends on a combination of new business and reduced overheads. The latter is in some ways the more important, for it can be deduced from Figure 5.7 that, whereas the 27 per cent increase in sales between 1974 and 1976 resulted in a 64 per cent increase in profit, a 47 per cent sales increase in 1976 and 1978 improved the profit margin by only 56 per cent. The profit-to-sales ratio tends to flatten out with increased sales. This is especially so when, to increase these sales, additional investment in extra staff or new plant is necessary. Against this background, some options open to Englass are as follows:

Option 1

To cease all manufacturing and become a purely marketing organization. There is some logic in this for, over the past twelve years, factoring has increased from 25 per cent to 55 per cent of total business. There are some attractions too. It would halve the labour force, reduce accommodation and energy costs, and remove the need for investment in new plant to expand production capability for new manufacturing business or automation, or both, in what is perhaps the maximum-risk area in terms of competition. Against this, this approach would increase commercial vulnerability and involve the disposal of some profitable and expanding activities.

Option 2

To concentrate on dispensers and ceramics, restricting manufacture to dispensers. This would avoid loss of the dispenser-manufacturing license, which is profitable and capable of expansion into such fields as injection-moulding. This could open up opportunities in areas such as can-closure, which would involve much-needed larger product numbers. This, in turn, would permit a restricted degree of automation and reduce labour-intensiveness. The existing rationalization of variation on a limited number of themes, such as the Falcon and the Stork dispensers, provides further opportunities for increase in product numbers by selling directly to consumers as well as to industry. Additionally, these increases in numbers might cause a downward adjustment of prices sufficient to make it feasible to incorporate dispensers into packaging and so increase market opportunities still further.

In the area of industrial glass this would be much more difficult. The most profitable and least competitive side of the business is in meeting specialized customer-requirements normally produced in small quantities. Moreover, changes in customer-demand have reduced the market for items capable of being manufactured with existing plants. A disadvantage of this approach could be that some customers are interested in industrial glass and dispenser or ceramic products. It is interesting to note that the acquisition of dispenser business came from the contact of supplying glass balls to a Danish pump manufacturer in 1964–5.

Option 3

To continue with all three current product areas but to devote expansion effort to the fields of dispensers and ceramics, retaining only profitable and customer-contact lines in the industrial glass area which are capable of being manufactured with existing plant resources. This would retain small profit areas and contact with manufacturers with an interest in more than one product area, without investment in automation or new plant in the area of industrial glass. It would, on the other hand, limit rationalization and retain an area of maximum difficulty in controlling overheads, e.g. costing for small numbers and high energy consumption.

Option 4

To attempt expansion into new business in all three areas. As far as industrial glass is concerned, this would involve new plant for glass over 3 inches in diameter, extensive research into the possibilities of orders up to ten times current maxima, with a measure of automation to ensure acceptable delivery lead times. Management would need to weigh carefully the balance between continuing market opportunities and investment before following this option but, if this option or option 3 were followed, it appears to be essential to prune out all small orders which are not highly profitable and regular, or making contacts for other products. The argument of the salesman that one should never stop producing something which can be sold has to be weighed against the overall effect on the efficiency of production.

These four options are not claimed to be the only possible choices nor are they mutually exclusive. They are intended merely to serve as guidelines when meeting a new challenge in a less favourable economic climate, requiring rationalization and pruning as well as a search for new business.

Appendix I Industrial Glass Products and Customers

Englass product	Customer	Customer product
Fire-polished bi-convex condenser lenses (44 mm diameter)	US market	V134 Viewlex filmstrip super viewer
Prismatic sight glasses	Rolls Royce	Spey Mk 201 by-pass jet
Aspheric condenser moulded by convex lens aluminized mirror and heat filter	Landward Engineering	Solaris projector
Ballotini beads	Dydale (Runcorn)	Hospital fluidized beds
Heat-resisting glass	Siemens	Domestic area
Moulded pyrex borosilicate glass disc	Fisons	Motorized dilutor to measure and eject concentrate diluent
Aspheric lens segments	Thorn	Airfield landing lights

Englass product	Customer	Customer product
Lynx sight-glasses manufactured from heat-resistant borosilicate glass	Westland Helicopters	Helicopter gear boxes
Lynx sight-glasses	Dowty	Undercarriage indicator in Tornado aircraft
Fire-polished lenses to replace more costly ground and polished lenses	Toolcraft Precision Engineering	Opera glasses
Fire-polished condenser lenses and aluminized mirrors	C & N Electrical Industries	Microfiche readers
Non-alkaline Ballotini beads	ICI	Grinding chemical compounds
Rod-moulded lenses	Oxley Developments Ltd	Solid-state indicator lens
Fire-polished aspheric lens	Vision Engineering Ltd	Eyepiece system for microscopes

Appendix II   Ceramics Products and Customers

Wade/Englass product	Customer	Customer product
UL 300 : 97.6 per cent alumina insulators	Colibri lighters	Insulators for cigarette lighter range
75 mm alumina ceramic caps	Aiton	Fitment to vertical tubes to combat wear in desalination plant
High-alumina insulation	Muller (W. Germany) — a company within the Philips Group	X-ray tubes
Vacuum-tight bonded metal and ceramic components	STC — Germany and Switzerland	For use in the electrical, electronic, chemical, and nuclear industries
Ceramic tubes	Cleveland Potash	Lining high-speed centrifuges coming into contact with abrasive elements such as found in potash, brine, and mined materials
Alumina ceramic protective sleeve	Connelly Equipment	Moisture-content in sand and fine aggregate used in concrete mixing plants
UL 500 Alumina skid tubes	Darlington Wire Mills	Guidance of carbon steel wire through furnace and acid-cleaning bath
Thick ceramic insulating rings machined to close tolerance	SRC	Accelerator tube in atom-smashing machine in nuclear facility at SRC's Daresbury (Cheshire) laboratory

Wade/Englass product	Customer	Customer product
High-precision tubes	Fischer Elektronik (West Germany)	Uniform pre-filtering of ex-radiation and X-ray tubes for dental surgery
Ceramic-plate capacitators	Ferranti	Oscillo-gyro for navigational systems
Electronic gun insulator	Cambridge Instruments	S 600 stereo-scan electronic microscope
Sections of accelerator tube in electron accelerator	Deutsches Elektronon — Synchrotron (DESY)	German nuclear ring main

Appendix III   Biographies of the Board

J.S. Abbott   Appointed to board in 1965 (sales director)

Born 1926. Member of Institute of Marketing. Engineering apprenticeship, Royal Navy 1941-4. Fleet Air Arm as maintenance engineer until 1956. British Empire Medal for service in Korean War. 1956-9 technical sales representative: Gulf Oil Corp. (GB) Ltd. Joined English Glass in 1961 as technical representative; then sales manager and subsequently sales director. Interests include sport, theatre, and musical appreciation.

R.W. Holmes   Appointed to board in 1959 (company secretary and
                          financial director)

Born 1926. Chartered Secretary: Associate 1958; Fellow 1969. War service: 1944-6 — Royal Navy, Telegraphist attached to Naval Intelligence Service. 1947-53 — group admin. clerk with British Road Services. Joined English Glass in 1953 as assistant company secretary and appointed secretary in 1954. Interests include gardening, photography, walking, theatre, music (particularly choral) and the arts, and various church activities.

T.J. Lawson   Appointed to board in 1959 (managing director)

Born 1931. Chartered Engineer. Obtained engineering degree at Loughborough and London Universities in 1954, with post-graduate apprenticeship at Westland Helicopter, Cowes (was Saunders-Roe). Joined English Glass in 1958 as development engineer; then technical director in 1959 and appointed managing director in 1970. Interests include swimming, sailing, photography, theatre, and opera.

H.A. Matthews   Appointed to board in 1965 (works director)

Born 1923. Trained as engineer. Joined RAF as wireless operator: 1941-6 served in South-East Asia. 1946-56 employed as engineer (fitter) and eventually planning engineer. Joined English Glass in 1956 as works manager. Interests include woodwork and various charity work.

<u>S. Perrin</u>   Appointed to board in 1962 (chairman)

Born 1915. Cost and Management Accountant: Associate 1938; Fellow 1953. Founder-member of British Institute of Management. Cost clerk 1929-31. Cost accountant 1934-46. Financial director and managing director 1946-62, when joined English Glass Company as deputy chairman. Chairman since January 1964.

Appendix IVA   Directors' Responsibilities

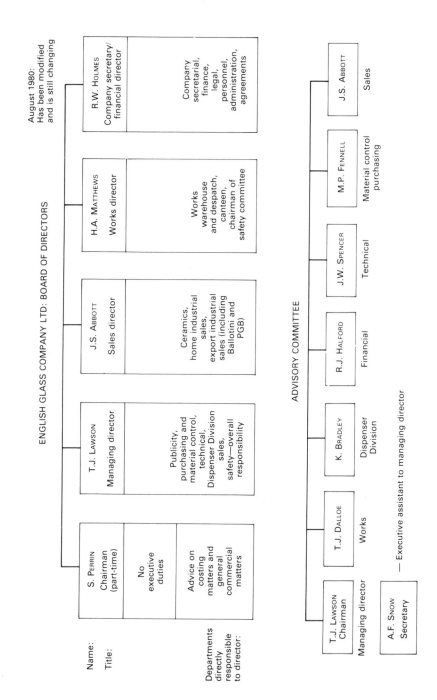

ENGLISH GLASS COMPANY LTD: BOARD OF DIRECTORS

August 1980:
Has been modified
and is still changing

Name:	S. PERRIN	T.J. LAWSON	J.S. ABBOTT	H.A. MATTHEWS	R.W. HOLMES
Title:	Chairman (part-time)	Managing director	Sales director	Works director	Company secretary/ financial director
	No executive duties				
Departments directly responsible to director:	Advice on costing matters and general commercial matters	Publicity, purchasing and material control, technical, Dispenser Division sales, safety—overall responsibility	Ceramics, home industrial sales, export industrial sales (including Ballotini and PGB)	Works warehouse and despatch, canteen, chairman of safety committee	Company secretarial, finance, legal, personnel, administration, agreements

ADVISORY COMMITTEE

T.J. LAWSON Chairman	T.J. DALLOE	K. BRADLEY	R.J. HALFORD	J.W. SPENCER	M.P. FENNELL	J.S. ABBOTT
Managing director	Works	Dispenser Division	Financial	Technical	Material control purchasing	Sales

A.F. SNOW Secretary	— Executive assistant to managing director

Appendix IVB Organization of Managing Director's Department

(August 1980: has been modified and is still changing)

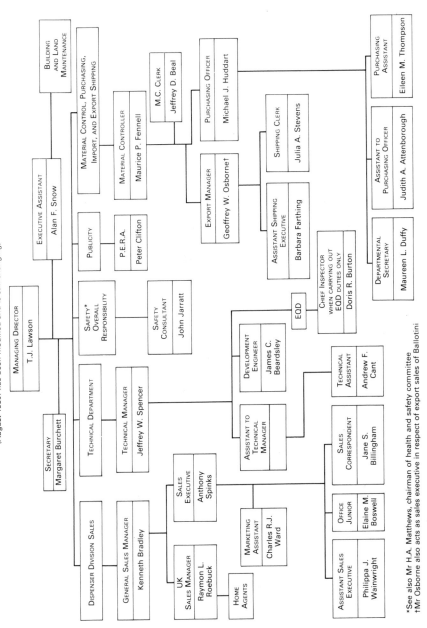

*See also Mr H.A. Matthews, chairman of health and safety committee
†Mr Osborne also acts as sales executive in respect of export sales of Ballotini

Appendix IVC   Organization of Sales Director's Department

(August 1980: has been modified and is still changing)

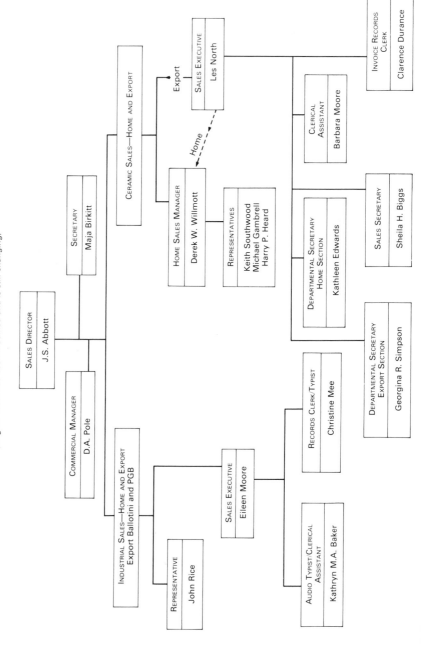

Appendix IVD Organization of Works Director's Department

(August 1980: has been modified and is still changing)

```
                    ┌─────────────────────┐
                    │  WORKS DIRECTOR     │
                    ├─────────────────────┤
                    │  H.A. Matthews      │
                    └─────────┬───────────┘
                              │
              ┌───────────────┴──────────────┐
              │                              │
   ┌──────────────────────┐       ┌──────────────────────┐
   │ CHAIRMAN OF HEALTH   │       │  WORKS MANAGER       │
   │        AND           │       ├──────────────────────┤
   │  SAFETY COMMITTEE    │       │  Terry Dalloe        │
   └──────────────────────┘       └──────────┬───────────┘
                                             │
                                  ┌──────────────────────┐
                                  │ PRODUCTION AND       │
                                  │ INDIRECT PERSONNEL   │
                                  └──────────────────────┘
```

Appendix IVE Organization of Company Secretary/Financial Director's Department

(August 1980: has been modified and is still changing)

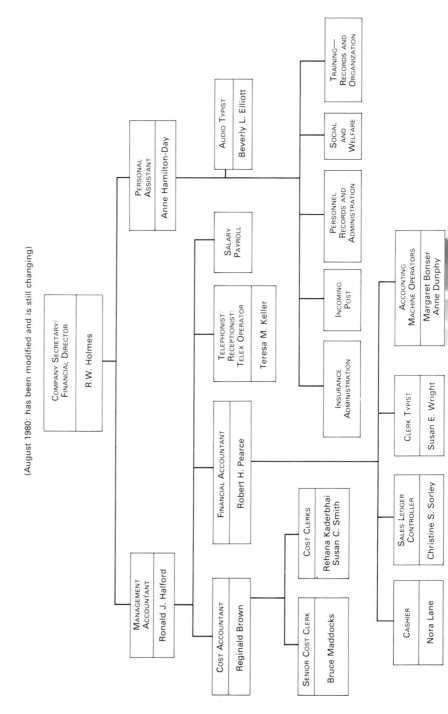

Appendix V Department of Employment Comparison

		30.06.80			30.12.79	
	M	F	T	M	F	T
Production employees	13	39	52	15	45	60
Engineering	4	—	4	3	—	3
Sub-total	17	39	56	18	45	63
Indirect works (Admin., Clerical, Warehouse, etc.)	9	6	15	9	6	15
	26	45	71	27	51	78
Directors	5	—	5	5	—	5
Directors' secretaries and assistants	1	4	5	1	4	5
Technical	5	—	5	5	—	5
Purchasing and shipping	4	4	8	4	5	9
Sales dispenser	4	2	6	4	2	6
Sales ceramics and industrial	6	7	13	7	7	14
Accounts	4	6	10	5	6	11
General Office	—	1	1	—	1	1
Total — Office	29	24	53	31	25	56
GRAND TOTAL	55	69	124	58	76	134

CHAPTER 6

Bestobell Mobrey Limited
(Case History No. 4)

An uncritical acceptance of the virtues of egalitarian society was one factor which encouraged permissive attitudes. As with many ideas which are sound in part it was abused, and gave rise to the neglect of many traditional virtues such as leadership and a belief in excellence. Leadership is difficult to define but can be recognized from its effect on people. The history portrays a company in which the actors are knowledgeable about the company's goals and committed to their attainment, and are therefore successful in meeting the challenge of new technologies and new ventures. The student may care to ponder whether this somewhat unusual company ethos is to be attributed to the leadership of the chief executive.

PART 1
Bestobell Mobrey Limited and its Direction

6.1 Company Background

Bestobell Mobrey Limited (BML) is a company long-established in the manufacture and sale of boiler-level control and fluid-level switches for industry. It is situated in Slough, Berkshire. In the 1950s it was a small supplier to the UK market; in the 1960s it expanded its activities by factoring solenoid valves and controls at home and overseas; in the 1970s it added more products—steam traps, flow meters, and other specialities—to its range. It now has a substantial international reputation in the field of electronics, electromechanical, and ultrasonic equipment and controls. A product range list and contracting services is shown in Figure 6.1.

BML is an autonomous subsidiary of Bestobell Limited, which is an international group in engineering, chemical, and consumer products with a turnover of £100m. The group has divisions covering the UK, continental Europe, North America, Africa, Australia, and South-east Asia. In other areas, it is strongly represented by agencies. In the last five years its operations have been disturbed by a poor profit record, by a change of chairman, by a restructuring exercise, and by a take-over bid. (Details of these are given in Appendix I.) Naturally, these events have had repercussions on Bestobell Mobrey. It has enjoyed the advantages of the group's widespread organization in expanding its own marketing and sales pro-

	Continental Europe only	Non-ferrous valves for steam, air, water, and associated services Cryogenic valves for low temperature Butterfly valves for gases, liquids, and free-flowing solids Boiler-mountings, level-gauges, and pipeline fittings
	Europe and United Kingdom	Boiler feed-water controls Liquid and dry material level alarms and controls Smoke-density indicators and alarms
PRODUCTS	United Kingdom	Valves (solenoid, etc.) for pneumatic and hydraulic control Controls for temperature, pressure, humidity Control equipment for environmental and industrial systems Electric/electronic/pneumatic control equipment and systems Hydraulic equipment and systems. Pumps Air-conditioning equipment Temperature controls, steam traps, and ancillary equipment for steam, hot water, and compressed air systems
CONTRACTING	United Kingdom	Design, installation, commissioning, and servicing of environmental and industrial control systems

Figure 6.1   Bestobell Mobrey Limited — products and contracting

grammes, but it has also suffered restrictions resulting from a policy of hiving-off successful developments to form separate companies within the group. In this way, BML has been parent to five or more specialist companies. Additionally, its management, both collectively and individually, has been considerably unsettled by the changing group situation, particularly by the take-over attempt. Great relief was expressed when it became known that the bid had failed. Throughout, however, BML has remained a major contributor to the group turnover and profits, although, as will be seen later, forecasts suggest a temporary drop during 1979–81.

6.2   Management Background

The company has always had a forward-looking attitude in its business objectives, as witnessed by the new products and general expansion which have continued for more than thirty years. In recent years, however, there has been a period of relying greatly on some of the established products which are now tending to become outmoded. With the retirement of the previous managing director after some eleven years in office the appointment of a successor in 1977 has heralded a new phase and provided a fresh stimulus for the management team.

The present managing director arrived with original experience as an engineer, then progressed through sales to marketing. As a result, he was convinced that marketing was the most important function in a company and that it should largely control all activities, at least up to the actual launching of a product. Even sales 'is really an arm of Marketing'. This thinking has accentuated an already market-orientated policy and has led to a definite dependence of all other departments. Nevertheless, he has inspired a general acceptance of his views, so that rivalry between departments has never become a serious problem. He has also generated an exceptional degree of commitment and enthusiasm for more sophisticated techniques throughout the company.

Because of his interest in marketing and a feeling that this area needed strengthening, he has himself become involved. He has recently appointed two MBA graduates in business studies from Cranfield School of Management and has considered the appointment of a marketing director. This led him into problems with the relative merits of his sales and marketing managers, and the decision finally made reflected his sensitivity to the feelings of his staff. The managing director showed wide general interests and was prepared to explore any relevant subject with which he was not familiar. He did, for example, attend an Ashridge Management College Workshop called in connection with the CEI project, and exhibited other enlightened characteristics which are rarely found in managing directors in the UK.

The marketing department is headed by Mr D.J. Morrow, who has been with the company for twenty-two years, of which twelve were spent travelling around the world servicing instruments. This long service has enabled him to see much of the development of the company, but it has produced a somewhat mixed view of the right way to proceed. He believes that the company has become top-heavy and is still only geared to relatively small operations. He appreciates the need to introduce modern methods but is reluctant to abandon those that served in earlier days. For instance, he supports the use of computers, but claims that 'ear to the ground' methods were more effective. As a result, much of the marketing philosophy is based on personal assessments derived from historical experience rather than from modern theory and analysis. One resultant strong belief was the need for marketing to be the arbiter of product policy, although this did not seem to be derived from the managing director's beliefs. He did, however, emphasize that sales staff are interested in immediate sales, whereas marketing must look forward to future demands and prepare the ground for sales. He expressed it as 'The salesman sells what he has got today; marketing sees that he has the right things to sell tomorrow'. His senior staff are termed product managers. They require a flair for engineering and must be able to discuss specifications, delivery dates, servicing support, and all other aspects of interest to the customers. They are also concerned with possible new products.

In summary, the marketing department's main problems are trying to assess what new products to introduce, how to do so without disrupting the sale of existing (but ageing) profitable products, and how to forecast requirements.

The sales department is five times larger than the marketing department: it is completely separate from it and employs more than 15 per cent of the company strength. This separation is unusual and is inherited from early company policy. It is organized under regional export sales managers who have individual salesmen operating in the regions through Bestobell subsidiary companies in some areas and through agencies elsewhere. The latter are not appointed at random but only where BML is able to service products.

The department is headed by the general sales manager, Mr J.F. Staal, who has been with the company for seven years. He is keen and energetic and anxious to rationalize the work of sales and marketing. He does not believe that separation of the departments has any advantages and appears to be undertaking some marketing functions. He claims, for example, that he is able to give guidance to other departments (particularly engineering and marketing) through his sales intelligence organization and that he does, in fact, do this in his monthly general sales committee meeting, which is attended by them. He offers sales and market predictions based on relatively simple deductions derived from charts, government data, discussion with sales staff, and even instinct! He discusses marketing negotiations with the marketing department but would like the marketing product managers to confine their interests to new products more than two years ahead, leaving more immediate developments in the hands of his own technical service managers. The latter are intended to be the experts on existing product lines but, under Mr Staal's direction, they are being encouraged to become more involved with new products. Here, he sees a division between the marketing requirements and expertise in the traditional electromechanical products and those for the new era of electronic devices which is approaching. The latter should be staffed by a new and separate group, which has both ideas and selling capabilities. Overall, he wants a strongly motivated team and believes it will be achieved by more integration of sales and marketing.

The other two departments which are important to the management team are engineering and production. The former is concerned in turning the marketing specifications and designs into practical products and ensuring that faults are rectified. A great deal of its work is concerned with customer service, which is to be expected since, by the nature of the business, a technical failure could shut down a complex process plant or threaten safety. 'Firefighting', nevertheless, has restricted the time devoted to research and development of new ideas. The chief engineer is Mr S. Wood, who has been with the company for eleven years and is now a member of the board of directors. He was originally a physicist, with interests and experience in magnetism and electronics. This he was able to apply to the company products, mainly in an evolutionary rather than an innovative way. His inclinations are towards the former and marketing situations have reinforced them. His departmental policy is 'First priority is the customer; second is production; and third is project planning'. He finds some frustration in the third aspect because advice on projects should come from marketing, but it is less than adequate. It fails to provide complete specifications which can define the work required; it is weak on financial justification; and it does not establish priorities.

The production department is responsible for all factory operations and associated functions such as stores, receipt and despatch, warehousing, and the toolroom. It employs 45 per cent of the company staff and includes the main factory at Slough; a minor site at Penn was included up until the time of its disposal in 1980. It is also expected to maintain a watching brief on Meterflow operations at Baldock. It has a number of modern tape-controlled machines, but has to handle too many special orders to give good results. The machine-utilization is high, but efficiency is low due to setting-up times. It has no resources for firefighting and anything more than minor departures from specifications are beyond its authority. In such cases it has to rely on engineering, with which it has a close liaison.

6.3   Company Situation in 1978

General

The initial reaction of 'The Management of Innovation' project team that the company was under a well-directed, professional, and enthusiastic management was confirmed during their first detailed investigation. It had just introduced, for the first time, a five-year strategic plan which took a frank look at its business situation, assessed its strengths and weaknesses, defined its future objectives and strategy, and laid down a timetable for action. This had been produced by corporate planning and discussion down the line of management and was, in fact, an example of the policy adopted by the company of consensus-based management. The company had a progressive personnel policy which used job descriptions and staff assessments and operated a training programme designed to encourage and retain skilled and qualified individuals at all levels and provide internal promotion prospects. Its total staff was just over 500 and the annual turnover was approximately 25 per cent, but a large proportion of this was accounted for by junior office staff.

In technical areas the company had recently introduced many modern techniques for the first time, including critical path planning, value engineering and analysis, brainstorming sessions, etc. It had a number of modern machine tools, e.g. CNC, NC, and transfer machines, but there were doubts as to their efficient use.

In marketing, it had been continually expanding its business for more than twenty-five years. Currently, its growth was by seeking new markets for existing products through small design changes which met international approval standards or other safety requirements. This had the advantage that it was not easily attacked by competitors, but the disadvantage that it resulted in too many 'specials' (i.e. modified products). A second major area of expansion was through licensing agreements so that new products augmented the home-designed ones.

The advantages that the widespread group organization might have been expected to produce were far from fully exploited. There was apparently no technology interchange, nor was any information circulated on market needs as seen by sales forces throughout the group. The main source of feedback to BML came from the five European companies which sold chiefly its products.

Business prospects

Although the initial review of company management indicated a praiseworthy attitude it also began to reveal some areas of weakness which could have serious future repercussions. BML had a first-class reputation for its basic engineering products, particularly founded upon their worldwide acceptance by reliability and safety-approval authorities. This was coupled with an ability and willingness by the company to modify designs as necessary to suit individual customers. This has tended to produce a dilemma—whether to aim for maximum sales through 'specials' or to achieve a high production efficiency with standard products. There is no doubt that, in the past, the 'specials' have received the major attention. As a result, much engineering effort has been diverted from its duties towards higher productivity, new developments, and other long-term operations in order to cope with the day-to-day firefighting emergencies which are constantly arising in the field with 'specials'. It has followed that efforts on behalf of tomorrow's customers are falling behind the attempts to satisfy those of today. There is thus considerable cause for friction between engineering and marketing departments. Fortunately, the personnel involved seem to have been conditioned by long exposure to the policy and by personal attributes to tolerate and live with it.

Another important fact to emerge was that some of the established products, which had served for many years, were becoming outmoded and had probably passed the point of maximum attraction for customers. Their past success had naturally brought attention from competitors who, particularly as patent protection ran out, were able to make improvements on the original designs and hence claim advantages. This was already beginning to result in a loss of markets and turnover. In addition, there was a situation of decreasing turnover of products manufactured under license or assembled and sold from outside sources. This was due, as noted earlier, to the group policy of transferring some successful products to other group companies, and also to the phasing-out of some license agreements and contract work. In 1977 the business from licensed products amounted to approximately 22 per cent of the total turnover, but this was expected to fall to 17 per cent in 1978 and, without replacement action, to only 14½ per cent by 1979.

Other factors have recently been recognized as tending to inhibit growth. Capital investment has been described as 'ludicrously small for many years'. As a result, there is now an urgent need for additional manufacturing space and a problem to overcome the inconvenience of two separate sites some 12 miles apart. There is also a legacy of environmentally sub-standard conditions in both existing factory and office accommodation.

Modernization is also needed in the manufacturing operations. There are inadequacies for example, in the manual system employed in stock control: it requires high staff numbers and unnecessary stock levels, and has a poor reaction time on scheduling changes. All this has a restricting effect on cash flow.

The combined result of these various factors was to reduce growth and, coupled with an anticipated downturn in general market conditions in 1978, to suggest even the possibility of a contraction of business and a falling level of profit unless compensating measures were introduced.

6.4   Company Plans

In the light of the self-analysis indicated in the previous section management had to face the question of how to overcome the weaknesses and to initiate new growth. It began by producing a five-year corporate plan which had two main objectives—first, the short-term arrest and reversal of the anticipated downturn and second, a longer-term stimulation of growth.

The immediate requirement was an increased volume of business which would follow improved productivity, better administration, and stronger marketing. This should be maintained into the longer term and improved by broadening the product scope, by widening the geographical and industrial coverage of the market, and by internal restructuring of the organization. Targets which these efforts were intended to achieve were set at a return on investment of 40 per cent by 1983, a return on sales of 15 per cent per annum, and a growth rate of 7 per cent per annum in real terms.

6.5   Product Range

If it is accepted that new products must be introduced from time to time in order to maintain and expand business some actions are necessary. These are primarily the responsibility of the marketing department, through which new requirements are established. It initiates such actions either directly through discussion with, or instructions to, other departments or, for more important issues, by putting them before the regular sales or development committees. Two levels of action have been exploited and a third is being encouraged. The first is internal minor developments, the second is external acquisitions, and the third is major innovations.

The first course has played a large part in the company's operations. It arises from customer requests for special products, from their suggestions or complaints about existing products, from the need to meet new operating regulations, or from similar causes. A great deal of engineering department effort has been absorbed in this work and has been reflected in greater product reliability and higher company reputation, but has not greatly enlarged the range of products.

The second course, that of acquisition, has been the major approach to extending the range. The company has looked for existing outside products which could supplement, or complement, its own and has then endeavoured to secure them by license agreement or by outright purchase. They have then been given, with the minimum possible additional development or modification, the backing and reputation of the company and introduced to the market.

A more elaborate approach on the same lines has been to seek out small firms with promising patents and developments, but without the strength to proceed to successful marketing on a large scale, and to acquire the complete company, lock, stock, and barrel. By using BML expertise to advise and guide, a new product can be added to the range in this way without excessive interference with the normal working of the company. Such a method was employed in 1977 with the acquisition of Meterflow Limited of Baldock. Its turbine flowmeter gave a useful addition to the product range and complemented the Sensall ultrasonic flowmeter previously acquired by license agreement. More background to this is given in Appendix II.

The third course, extending the product range by in-house innovation or sponsored institutional research, has not been greatly exploited. It has been advocated by the development committee, but has still to overcome the longstanding attitudes of staff who have been conditioned to the other approaches.

Development, acquisition, or innovation?

The company has built up its product range through the years by recognizing and acquiring basically sound ideas and improving them. This served it well during the period when electromechanical technology was supreme, but now this is being overtaken by new technologies—electronics, computers, microprocessors, and other devices. A question immediately springs to mind—how does this affect the company's future? Management must answer this, and upon the answer will depend the company's long-term development plans.

The marketing manager expressed the position clearly when he said: 'We must choose between evolution and innovation and this is the problem to be solved within the next five years.' Strongly advocating innovation has been Dr Hugh Conway, consultant to the company and, at one time, the first chairman of the CEI Committee on Creativity and Innovation. His advice was that new ideas be stimulated by think-tank sessions, by setting up a research team, and by other unconventional means. The managing director has accepted that something on these lines is desirable and has proposed that a 'mad money' fund should be established for such investigations. He has also encouraged university research, sponsored by the company, and believes that this acts as an external stimulant for staff. At the same time, he realises that the past strength of the company has been in other directions—'We are better at doing and modifying than at initiating'.

The chief engineer has pointed out that the mechanical work is largely evolutionary, while the electronic side is innovative. His inclination is towards mechanical devices, although he does not dismiss electronics altogether. He is in favour of acquiring new ideas from outside, but warns that a great deal more in-house study should be carried out than is presently undertaken in order to judge adequately the potential of such possible acquisitions.

The marketing manager is strongly in favour of acquisitions, since 'We then have a potential input from all engineers in Great Britain. This should be cheaper and more certain than supporting a large in-house research team. The only problem is in identifying the winners before the competitors do'.

At present, winners seem very much a matter of chance. In recent cases they have been produced mainly because of the persistence and judgement of individuals who have followed up inventions. The Doppler flowmeters and ultrasonic level indicators and the latest level-switch (to be discussed later) were all developed initially under such conditions.

Thus the situation is far from clear-cut and management appears very uncertain as to the right action. There is a leaning towards the new and innovative ideas, but a hesitation to abandon the old and well-tried methods of product development.

Electromechanics or microelectronics?

Experience has shown that almost all new requirements can be met by electromechanical means and, for many years, the company has been geared to this, both in staff experience and in workshop equipment. Modern technology has now indicated that electronic devices can supplement, or even supplant, many of the older techniques, and this poses the question as to whether, or how, to adopt such a change.

In favour of solid-state development it can be claimed that moving parts are eliminated, no maintenance is required, non-penetrative designs can be used, and that complex functions can be monitored and controlled. On the other hand, the electromechanical devices are also capable of controlling many functions and, in some areas, more certainly and more effectively. In the field of traditional boilers, for example, they can operate in high temperature and pressure environments, their reliability is very high, and confidence is increased by the ease with which mechanical checks can be made. They do not rely on an external power supply to function and only need one pair of wires to provide direct signal or control operation. The electronic black boxes need more complex electrical connections and servosystems to support them, they need to be self-checking (which may involve electromagnetic contacts with limited life), or, in some cases, where very high reliability is essential (e.g. nuclear plant), they may have to be fitted in duplicate or even triplicate. Their value is, however, enhanced when they can be fitted without requiring cut-outs or holes in pressure vessels or pipes, so eliminating fracture risks. Generally, however, they are not so easily adapted to meet safety authority requirements or EEC regulations as are the electromechanical devices.

From the customer's point of view some firms will find attractions in advanced technology as a way to improve their own image, but others will have a strong distrust which will be difficult to overcome. They prefer hardware that is understood by their staff (e.g. boilermen) and which can be serviced *in situ*.

From the company point of view any major change in its products will have far-reaching effects. An increase in electronics and decrease in mechanics would require a change in staff expertise, would reduce the utilization of the machine shops, and would be dependent upon skilled and scarce electronics labour. The company's present superiority would be severely challenged, as there are many potential competitors in electronics. Many of these are small firms who can operate on small capital, can cut corners in production, and would eat into the market.

Within the company there has been a general view that the basic engineering products at present cannot continue to sustain turnover but, equally, there is a reluctance to accept that this is inevitable. There is a somewhat less general opinion that electronic products must replace them. It has been realized that the company is already involved in this field with its ultrasonic level indicators and Doppler flowmeters, but that these products represent less than 12 per cent of the turnover. Additionally, they are regarded as no more than alternatives to the normal magnet-controlled level switches and turbine flowmeters. Nevertheless, this business could indicate the least disruptive method of establishing solid-state technology as a major part of the company's development. A phased change over a period would enable stocks of old products to be run down, a build-up of technical know-how, and a marketing effort amongst customers.

6.6 Marketing Strategy

The traditional markets for company products have been the power and process industries, particularly in providing fluid-level information and in fluid flow-rates. More recently, the chemical industry has become an important customer for special versions of the company's products, and new openings are occurring in North Sea oilfields and in the nuclear industry. These need study to find how their requirements can be met by existing, modified, or new company products. A start on this has been made by constructing matrices of products marketed against the parameters which customers wish to measure or control.

Arising from these, it has become evident that higher levels of safety and reliability will have to be established in qualifying for acceptance by some approval authorities. To meet these, more enginering staff and equipment, especially electronic engineers and test facilities, need to be recruited and a new quality assurance programme introduced. A development timetable for those products which are to be available in the two-year period from 1978 has also been drawn up, emphasizing particularly the dates when engineering production and marketing should commence or be completed. To these, management has decided that the acquisition of a further small company, possibly with expertise and products outside the normal BML range, would be advantageous. In this way, the change towards new techniques and the need to extend the product range could both be met without too serious an interference with other operations.

Beyond the immediate future the matrices reveal the absence of an instrument for detection of explosive gas-concentrations and a requirement for a more advanced density-concentration monitor. Potential markets for these have been identified in the petrochemical industry and in various industries concerned with sludge and sewage-effluent control and disposal. The former already exists in the USA and plans are now in hand to seek suitable agency or manufacturing licenses. The latter exists as a simple on–off indicator and is marketed by the company among its Sensall ultrasonic range of instruments. Continuous indication is a desirable extension and such development is scheduled within the company.

In the longer term (horizon planning) it is planned to continue and extend the monitoring of research activities at universities and government establishments to identify new ideas which might materialize into saleable industrial products over the next decade. Collaboration will be offered in suitable cases. Particular fields with future potential include the use of lasers, microprocessors, and ultrasonics.

Factored products extend the company range without involving a major demand on internal resources and generally provide a sufficiently attractive return to justify active selling effort. The current position is less than satisfactory, however, because of the losses referred to earlier, and demands special attention. The first step must be an expansion of the existing products. Some of these have established a good reputation but have not been greatly exploited. Now a detailed marketing plan is required to form the base for an aggressive promotional campaign. The second step will be to consider further products to replace those which have been withdrawn. Some of these were subject to restrictive agreements which limited sales to UK territory. The effort now will be to replace them with much more extensive franchises, which will allow their exploitation through the very extensive BML overseas agency network.

With these developments in products, marketing will have, as a major objective, an increase in the company's share of the overall market. To assess the situation it has made analyses of both manufactured and factored products for home and world markets, of the present share, and of that in 1983. It has also listed competitors and their market share. From the results it is clear that the greatest scope for growth at home is in the relatively new ultrasonic controls and in solenoid valves but, overseas, there is still a useful growth, possibly for traditional types of products. There must therefore be a big drive to export to world markets, and charts have been drawn up to indicate the most efficient disposition of effort. It will include the use of a mobile exhibit in the USA and a separate one in the UK and Europe, use of sales task forces, special sales conferences, and launching publicity for new products. The effort will not be confined to Europe and America but will extend to the Middle East and Comecon countries and will include the exploitation of new territories and the appointment of suitable agents wherever this can be justified.

6.7 Internal Organization

As part of the management plan for revitalizing the company it laid down that there should be 'a continuous updating and modernization of internal operations'. The plan envisaged a 25 per cent growth in manufactured product sales by 1983 and, at the same time, required higher productivity, lower overheads, and a better return on capital employed.

Immediate steps were taken to meet these requirements. The first was clearly to provide for the increased production and the second to do so with greater efficiency. An investment programme was drawn up for providing additional tools and replacing obsolete ones, together with plans for additional floor space and a reorganization of the work flow. The total capital expenditure over five years was assessed at nearly £1.4m, of which the greater part would be used in the first two years. This was a bold decision in the light of the previous reluctance to commit any funds to capital development. It was expected, however, to result in higher productivity and to reduce the demand for additional operatives to meet the planned expansion by a third. In cash terms, this should give a saving of nearly £50 000 per annum.

The second step was to deal with some of the paperwork systems. It had already been recognized that the existing manual systems of stock management, sales and orders control, invoicing, accounting, work scheduling, etc. were not as efficient as they might be. During 1977, the group computer had been programmed to deal with BML stock records and this experience indicated how further administrative areas could be planned to reduce manual procedures. A systems action plan was prepared and it was anticipated that, over two years, its implementation should yield benefits from reduced stock-holdings, improved cash flow, lower administration costs, and better management information. The immediate aim was to be an improvement of 10 per cent in two years but, ultimately, the result should be greater management control over all associated production and administrative functions so that the planned expansion could be handled without any additional staff.

After allowing for the improvements expected from the foregoing plans it was still apparent that restraints of factory and office space at the main works would limit the growth there. To overcome this, management has explored the possibilities of setting up self-contained satellite establishments where a group of employees would be able to develop, manufacture, and take full responsibility for some particular product or technology. In this way the group would identify itself with the product and build up an expertise and affinity with its product. Meterflow was considered a good example of this arrangement. It has not been possible to develop this structure very far, but a start has been made recently by establishing three business groups as separate profit centres. They are responsible, respectively, for level switches and associated products, electronic devices such as ultrasonic levels, and new products. Each is headed by a senior management member and they roughly divide the business between traditional products, factored products, and new products.

PART 2
Bestobell Mobrey Limited—Level-Switches

This study has been divided into two parts, which can be examined independently. Part 1 reviewed the history of Bestobell Mobrey Limited and the recent introduction of new thinking and modern concepts within the company in a broad perspective. Part 2 now considers how the philosophy has been absorbed at middle management levels and applied to one particular section of its operation.

At this point, it is appropriate to make reference to short biographies of the chief personnel (Figure 6.2), and take a look at the organization chart

Figure 6.2   Biographies

G.G. Woodhead   (Chairman/managing director)

Mr Woodhead is both a mechanical and civil engineer. Before becoming managing director of Bestobell Seals in 1972, he had been director of marketing in a precision engineering group and in the aerospace industry. He became head of Bestobell Mobrey in 1976, and was appointed group manager of the newly formed controls and instrumentation group in July 1979.

S. Wood   (Technical director)

Mr Wood graduated from Aberdeen University in 1957 with an Honours degree in physics. He joined the Marconi Company on a two-year graduate apprenticeship and spent the next eight years in the research and development laboratories investigating the use of magnetic materials in non-reciprocal devices. He joined Mobrey in 1967 as development manager, was promoted to chief engineer in 1972 and was appointed to the board in 1974. He now has a dual role of technical director and business unit manager for electronic products.

J.F. Staal (Sales director)

Mr Staal's nationality is Dutch, and his qualifications are equivalent to BSc in mechanical engineering through the Higher Technical College, Amsterdam, and the Technical University, Delft. Previous experience included drawing office estimating work with the Express Lift Company and selling experience with Aerofill Limited, Filling Machinery. Since joining Mobrey he has been involved in selling in export markets and product management, which led to export sales manager, then general sales manager, and finally sales director with Bestobell Mobrey Limited.

D.J. Morrow (Marketing manager)

Mr Morrow spent two years' National Service in the RAF before joining Mobrey in 1956. He started as service maintenance administrator before taking the opportunity of going to Australia with the Bestobell Group as assistant technical manager in 1960. He was promoted to technical manager in 1963, and returned to Mobrey, Slough, as marketing manager in 1972. On the return trip from Australia to Slough he stopped off in Toronto, Canada, where he was general manager for two years. He now has a dual role in the company as marketing manager and business unit manager for factored products.

J.W. Davison (Product manager (Level-Switches))

Aged 34. Secondary education followed by HNC with endorsements. Mr Davison started work with his father, was involved with precision toolmaking and prototype manufacture, and managed this business for three years. He joined Wilkinson Sword in 1967 as laboratory technician for five months, then joined BML that year as junior draughtsman. He had various promotions within the drawing office to section leader, plus a short time as development engineer, until 1977 when he transferred to become product manager (level switches).

P.M. Downing (Mechanical development manager)

Aged 32. Dr Downing went on to obtain an Honours degree in mechanical engineering and a doctorate in fluid flow measurement, both at the University of Surrey. First introduced to industry during vacation employment from school and, subsequently, university, working as a draughtsman and later as a fitter/machinist. He joined BML in 1977. Previous experience includes development with Rank-Xerox working on new-generation copiers and peripheral devices.

N. West (Senior development engineer)

Aged 44, HNC mechanical engineering. Mr West joined BML in 1953, left to carry out National Service, and returned in 1956. He gained experience within most areas of the company and joined the development department in 1963 as a test engineer. He was promoted to development engineer in 1968, became senior development engineer in 1977, and therefore has intimate knowledge of many of the company products.

116

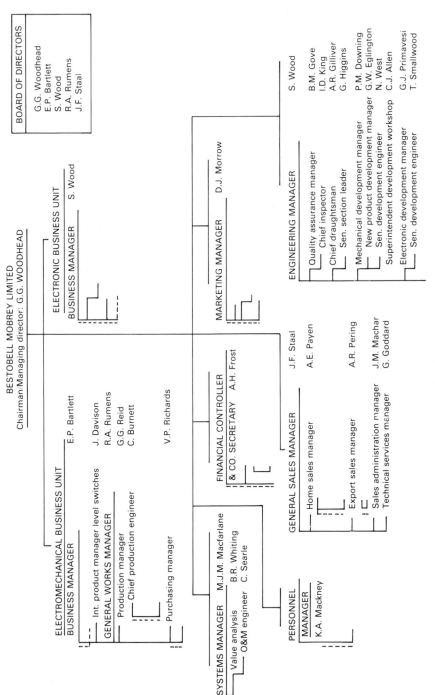

Figure 6.3 Organization chart

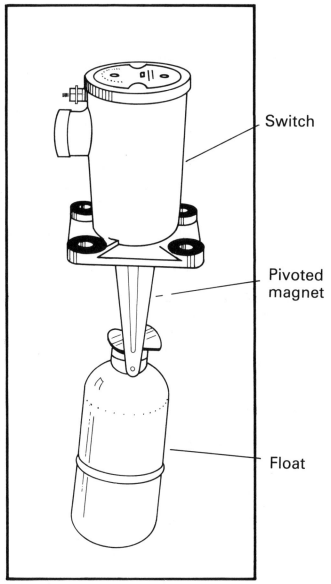

Switch

Pivoted
magnet

Float

Figure 6.4   Mobrey magnetic level-switch

(Figure 6.3) in order that the reader may imbibe something of the
company atmosphere.

6.8   Background

Switches operated by small magnets in conjunction with floats had been the
basis of the company range of fluid-level products for nearly thirty years.
Figure 6.4 shows the popular Mobrey level-switch. The number of different

arrangements and devices to which they were applied had grown enormously by catering for different industries and different operating specifications. The latter became increasingly rigorous and resulted in many variants of the basic design. For example, switch contacts were made from beryllium/copper for operating temperatures up to 210°C, from inconel above 210°C, and plated with gold or platinum/rhodium in some cases where voltages varied from standard. Added to these were alternative materials for switch bodies and other design changes for special purposes. Altogether there were thirty-eight variants of the standard switch, although many were only supplied in small quantities. Other design features which were producing criticism from the users included the four-contact arrangement, which was limiting the combinations of circuits which could be opened or closed. Users also found that the wiring terminations were not easy to work on.

A more important cause for concern, however, was the increasing severity of the vibration requirements which, in some cases, called for contacts to withstand breaking forces up to 4 G or more over a broad range of frequencies. Unfortunately the existing instruments could only withstand 1.8 G and had a resonant frequency almost in the middle of the range. This explained the observed cases of momentary opening of circuits. Although this might be only of microsecond duration, it was unsatisfactory and might lead ultimately to a serious loss of business.

The switches provided a substantial part of the company turnover and held an 80 per cent share of the home market, as well as 20 per cent of that overseas. Competitors had been able to make some inroads by copying the design and, in one case, by using a closely similar trade name and address. Generally, because they started without the restrictions imposed by the original design they had been able to introduce one or two improvements on which to base their competition. The net result had been a drop in sales of 5 per cent per annum over the previous five years.

In 1972 the company tried to come up with a new switch which would meet all 'authority approvals' and be compatible with the horizontal and vertical modes. This was the situation at the end of 1977 when management realized that some action must be taken quickly if the level-switch business was not to be heavily eroded. Early in 1978, Dr Peter Downing, the recently appointed engineering manager, advised that development work should be restricted to the horizontal switch, otherwise an acceptable cost target could not be achieved and, in conjunction with marketing, he gained acceptance for this view. Fortunately, the basis of a solution was already available.

6.9 New Switch Origins

The development engineer responsible for level-switches, Mr Norman West, spent a lot of his time investigating cases of switch-malfunctioning and finding solutions to their problems. He had an analytical mind, was experienced in design, and was an untiring worker. Mr West had been greatly encouraged by the 'new look' policy introduced by the managing director, which he described as 'exactly what the company needed' and which had improved morale at all levels. He also commented on the timely support given to him by his immediate manager, Dr Downing.

Before this improvement, however, he had received complaints of two separate incidents in which level-switches failed to operate in appropriate conditions. Both switches, when tested, seemed satifactory but were replaced. Nevertheless, he concluded that there had been some malfunctioning of the contacts and that an increase in pressure between them was the way to avoid repetition. Intervention of more immediate problems prevented active work on the problem and only spare-time effort was possible. In spite of this, he produced a design to improve the contact pressure but, although there was still no official enthusiasm for his ideas, Dr Downing asked him to continue his work, whenever other commitments would allow and, in particular, to pay special heed to achieving high rigidity. By 1977, his persistence and skill had produced a prototype switch which seemed likely to overcome the contact troubles and promised other advantages. This work was reported, but no action was taken and the model lay on his desk for some months. This lack of interest was typical of the relatively shortsighted policy at that time and, only later, when the new management attitudes began to permeate through the departments and the full implications of the trouble arising in the switch market began to emerge, was it realized that the new design might provide a fresh lease of life for the product. The marketing department then began to think about the general applicability of the idea and also its value in rationalizing the various forms of switch. Finally, by late 1978 as mentioned above, senior management was persuaded that the design should be developed and its potentialities exploited.

6.10 Project Team

Once approval had been given by senior management to proceed there was immediate pressure upon the technical departments to develop the design into a marketable product. It was clear that this could not be done without direction and co-ordination of effort between departments. The engineering development department was probably most involved, although marketing had a major interest, and at various stages, the drawing office, the production department, and quality control had important roles. To ensure that they were all able to contribute advice and help a monthly liaison meeting was held throughout the development. This meeting co-ordinated activities and acted as a project team and avoided the rigidity of a formal committee with a permanent chairman and fixed membership. It was always referred to as the 'DS77 Push Rod Switch team', and was very flexible in its membership, including the post of chairman. This was first held by the engineering development manager, Dr Peter Downing, and later, at an appropriate stage, passed to the product marketing manager for level-switches, Mr John

Davison, and, finally, to Mr C.A. Burnett, chief production engineer. The meeting had no appointed secretary, but minutes were prepared and actions recorded by the chairman. The informal way in which the change of chairmanship took place reflected the distinctive nature of the co-operativeness of middle management.

The members of the team were all comparatively young (30s and 40s) and proved themselves keen and able. Their discussions were naturally biased by departmental interests, but were almost always constructive and never obstructive. Some of the credit for this willingness to compromise and to cut corners when necessary must be given to the chairmen, all of whom influenced their meetings by an ability to be objective and decisive, although, on two occasions, when their concentration and energy were lower than normal, it was soon apparent that the meetings deteriorated and lost their usefulness. On such occasions, long discussions might occur on minor points and end without any conclusion or useful decision.

Such points, however, only emphasized the value of the meetings, which helped to build up a remarkable commitment by a body of employees to achieve success. When asked whether other products had received similar treatment, both Dr Downing and Mr Davison replied 'Good gracious no!'. This is the biggest development in the history of the company and, in any case, would not have been possible under the previous management'.

6.11   Feasibility Phase

Although, in principle, the new switch had great promise the practical aspects had not been investigated in detail when authority to proceed was given by the executive management. These required a logical step-by-step assessment of the project and became the responsibility of the middle management team.

The first step was to achieve a broad agreement for the switch specification, which was prepared initially by Dr Downing in collaboration with Mr Davison. It was critically examined by the project team at its first meeting and then formed the basis of all future work. Not unexpectedly, a considerable input came from the quality control manager. It was to incorporate the new 'push-rod' operating concept (see Appendix III) and to be techically superior to both its predecessor and the products of competitors in contact force, electrical rating, operating temperature range, vibration range, and versatility. The latter would be achieved by rationalization of the many existing variants into five types. The first would be a direct replacement of the existing four-contact single-pole single-throw switch, the second, a new six-contact double-pole double-throw switch, the third, a similar six-contact switch but hermetically sealed. More information of the requirements is contained in Appendix III. Types 4 and 5 were modifications of types 1 and 2 that enabled them to cope with special environments. Gold contacts were used and stainless steel in the bearings replaced phosphor-bronze.

The second step was to undertake a feasibility programme which would include prototype design, testing, and costing. This was drawn up in the form of a PERT chart, again prepared by Dr Downing. He had introduced this system of control to the company and was now able to direct the project team in its use. The chart was accepted without question and, although difficulties arose in maintaining the scheduled dates during the initial phases, these were mainly caused by the conflicting demands of other urgent work overriding the programme. This highlighted a weakness in management, which accepted the priorities mentioned above (page 106) but largely ignored the consequences. PERT charts were constructed for the development and pre-production stages of the four-contact, six-contact, and six-contact hermetically sealed push-rod, and enabled the many and varied activities to be effectively controlled. They served as the agenda for meetings and directed attention to necessary courses of action. Efforts to establish a firmer base for the DS 77 project were made by calling a special time-scale review meeting attended by the managing director, the chief engineer, and the marketing manager. This gave some help by relaxing limitations on common design for all three forms of switch but did not lay down any definite priority.

Also in the early stages the newly formed value engineering section was given an opportunity to review the requirements and functions of the new switch and the problems associated with the old one. In doing this, it also speculated upon alternative ideas to meet the needs. A brainstorming session produced a number of concepts which are outlined in Appendix IV, but it was clear that DS 77 was too far advanced to justify a fundamentally new approach. During later costing exercises, however, value analysis was applied to the design and a cost-function matrix revealed that the push-rod assembly was accounting for 30 per cent of the switch cost.

The feasibility phase was concluded by March 1979, when the project team presented a review to senior managers. This review was encouraging technically, but indicated high costs for the switch. Management accepted the review and agreed that the next phase of development should proceed while putting extra effort into the study of cost-reduction.

6.12   Development Phase

The second phase aimed at carrying the project to the preproduction stage, mainly by further development and test programmes which would establish the switch approval with safety and other authorities. Quite early in this phase, the major responsibility for the project team began to shift from engineering to other departments and this was accompanied by a change of the co-ordination meeting chairmanship from Dr Downing to Mr Davison since marketing was primarily involved. This was done smoothly and without problems, Peter Downing continuing to attend the meeting and provide an engineering input.

One of the notable aspects of this phase was the endeavour to follow the PERT programme and the attitude towards slippages from it. This seemed, at times, somewhat ambivalent. Dr Downing regarded any delays resulting from design or development causes as project failures, which should be remedied as far as possible. On the other hand, slippages resulting from project effort being diverted to other work were beyond his control and accepted as a consequence of inevitable firefighting activities for Bestobell's customers. His view was that project work and normal sales and work queries could not be segregated within the department without discontent and falling interest in engineers allocated wholly to the routine work. (Note: towards the end of this study, much of the firefighting had been transferred to production engineering and Dr Downing estimated that in his department it had been reduced from 40–60 per cent to 10 per cent.)

The contrast between this situation and the endeavours to recover lost ground, or reduce exceptionally long programme times in other cases, was remarkable. The project team was always willing to attempt to speed up critical path steps in order to recover lost time and this co-operation continued under Mr Davison's chairmanship. It was illustrated by the efforts to improve the preproduction PERT programme, which indicated an unacceptably distant completion date. Actions to expedite this were suggested by team members and agreed by departments: for example, the drawing office arranged for subcontracting of drawing work, and the production department agreed to by-pass production control and production planning for critical components in ordering materials and, in conjunction with the development staff, was also able to approach suppliers with modified requirements, so that several months were saved on deliveries.

Although the emphasis was moving towards production and marketing, new development aspects continued to be raised. Only at this relatively late stage was a proposal made to improve the switch housing by changing from the existing standard body (SO 1) to another which had been used in special applications only (S 196) in spite of the value engineering meeting which originally made the suggestion (see Appendix IV) during the feasibility phase. Study showed that the latter had several advantages including, particularly, easier terminal wiring access. It did, however, introduce questions of extra tooling cost. This change was one of several, e.g. an updating of the float design, which were considered late in the programme.

A more fundamental problem appeared in the vibration test programme. A resonant condition was found in the middle of the frequency range, which caused the switch to open at force levels below the 4 G specification required by some authorities. It was ultimately traced to a resonance in the test apparatus, which occurred at 47 cycles and could be altered by adding mass. This had, of course, been one of the problems with the original switch and now raised the question 'Was the original better than had been thought?' Although the aberration at 47 cycles could now be accounted for, the 'G'/frequency characteristic of the push-rod switch was a very significant improvement.

This situation arose during the period of the BTR take-over bid (see Part 1) and the team was uncertain of the company's attitude in the light of its probable need to show improved short-term profits. It felt the need to receive further endorsement of the project by the executive and therefore it called for another presentation. This was mainly concerned with the financial justification. Not only was it necessary to allocate money for tooling but the DS 77 switch itself was proving to be more costly than the original switch. A preliminary estimate some months earlier had shown a unit cost of £9.95, compared with the G 350 cost of £5.11. It had been expected that costs could be reduced and, in fact, it was possible to get the figure down, first to £7.20 and, later, to £6.50, by increasing tooling costs from £25 000 to £37 000. Even these figures were received with some consternation by the marketing manager but, after balancing against the other advantages, he accepted that they were a basis for further financial authorization. The project was then allowed to proceed with a requirement for a later justification of costs.

Further attacks on cost were made, largely by Mr Norman West (development engineer, who proposed the new design). He made a series of visits to suppliers of parts and materials and was able to obtain agreements on reduced costs by minor design compromises which, ultimately, brought the unit cost down to £5.46 with a total tooling cost of £44 000.

The team also suggested other cost-reducing possibilities. Value analysis was one and the use of second sources of supply and the separate procurement of dies were others. A policy of more aggressive negotiation of prices with suppliers was adopted. There were problems associated with all these ideas and their cost-effectiveness was considered not to be worthwhile at this stage. The justification should, therefore, not rely on cost alone, but more strongly on marketing considerations such as the benefits of moving up-market with the new switch and the advantages of rationalization.

In parallel with the work on costs the final stages of development were completed and planning for the preproduction phase was actively pursued. This involved the determination of pilot batch sizes, the organization of field trials (a necessary and important prelude to the market introduction), and the scheduling of production quantities and timing. Before, however, this phase could commence a final justification report had to be prepared and approved by the executive. It was compiled by the team leaders with help from other departments and gave a comprehensive case for continued action (see Appendix III). It was presented very effectively with visual aids and created an excellent impression at a meeting in December 1979. It was accepted and authority given to proceed, subject to three conditions:

(1) The work cost be monitored every two months;
(2) Action be taken to imbue the sales force with enthusiasm; and
(3) A plan be produced to fit into production, giving stock balance between new and old.

6.13   Preproduction Phase

It was planned that this phase should be completed for all three basic versions of the switch and that they should all have started production before the end of 1980. Preproduction was estimated to require a total of 450 units which would be used on performance tests and in field trials. While these were proceeding, preparation for production would be in hand with an ultimate target of 60 000 units per annum. This should be reached over a six-month 'working-up' period and an equal 'running-down' period for the old switches. Production should start at 10 per cent of the full rate and rise to 1200 units per week.

Although, at this stage, the production department was becoming much more involved, it continued to play a subsidiary role to marketing and sales. To help in ensuring its full awareness and planning co-operation a new separate 'Introduction to Production' meeting was instituted. This was attended only by those directly concerned with production, together with Peter Downing, and John Davison who still retained his role as chairman and meeting reporter. Such a meeting had never been held before but it quickly illustrated its value. Almost immediately it became clear that all those present were not familiar with the up-to-date design position. The new switch body (see section 6.12), for example, introduced changes of which they were unaware. It also highlighted the time delay between ordering and receiving sophisticated tooling, it considered the possibilities of 'knife and fork' methods of preproduction, and it emphasized the risks of ordering high capital-cost machines while any development doubts remained. These considerations were, in fact, set aside by Peter Downing's confident 'I have no doubts'.

Although this meeting proved so helpful and provided a remarkable example of the commitment of all concerned to achieve success, it was only repeated once or twice before its actions and progress comments were incorporated into the main project meetings. In some respects this was regrettable, as there appeared to emerge some departmental feelings which had not been present during the earlier phases. It may have been merely that the chairman was overburdened and did not have his normal adroitness in handling the meetings. On the other hand, the production representatives may have felt that their interests were not receiving sufficient attention. However, during this phase Mr Davison passed the chairmanship over to Mr Burnett, as mentioned above. Another factor may have been the increasing number of persons attending the meetings. Later meetings were reduced to those who had a definite contribution to make.

At this stage, the sales department began to take an active part in the development, in line with the executive directive. Meetings to present the new switch to sales staff and instruct them on its merits were organized and further meetings to induct Continental associates were also scheduled. Decisions on how to change over and which versions could replace old switches were made and any consequent actions were agreed.

6.14 Conclusions

At this time (August 1980) the preproduction PERT programme is still proceeding and the critical path is well on target. Numerous small difficulties are arising, but solutions are being found without loss of time. The project is therefore going ahead at this stage very much as planned.

Appendix 1 Bestobell Group

Bestobell is an international group operating in the UK, continental Europe, Australia, South and central Africa, North America, and South-east Asia. The group's activities are in three related areas of engineering and in chemical and consumer products. The engineering areas cover manufacture of equipment for transmission and control of fluids and thermal and acoustic insulation, merchanting of equipment which is biased towards the manufacturing products and contracting for the installation of insulation and air conditioning. It has been described by the *Financial Times* as a 'mini-conglomerate' with 'such diversification in products, technology and geographic markets as to put its prospects practically beyond the comprehension of all but the most assiduous City analyst'.

Its sales have risen steadily throughout the 1970s but its profits have failed to keep in line, as shown in Figure 6.5. It was recognized that these unsatisfactory results required changes and, in June 1979, the part-time non-executive chairman since 1969, Sir Humphrey Browne, stepped down in favour of Mr A.B. Marshall who became full-time executive chairman. He had previously been managing director of the P. and O. Group and was a well-known figure. He was expected to restructure the group management 'to restore the level of performance and obtain a resumption of growth'.

Year	Sales (£'000)	Profit (£'000)
1975	63 008	5691
1976	76 874	5335
1977	85 615	5490
1978	95 496	4900

Figure 6.5

Within ten days of this change, however, BTR, an industrial conglomerate, made a take-over bid which was strongly resisted by Bestobell. The bid finally failed in September 1979 and this was ascribed to a mixture of skill, faith, and luck. The new chairman, in conjunction with the managing director, Dr D. Spencer, was able to forecast increased profits, to claim reorganization in progress, and to persuade the major shareholder to stand by the group. He was also helped by the City's knowledge of his reputation.

The bid may also have proved a blessing in disguise, because both managers and workforce rallied round the chairman, so enabling him to expedite the restructuring process without serious internal opposition. This new structure, shown in Figure 6.6, consisted of four business groups within the UK and two regional groups covering Africa and Australasia and South-east Asia.

The group managers now meet as a management committee each month and form a coherent management structure with clear lines of profit responsibility. There is also a much clearer description of the group's activities and this is indicated succinctly in the new letter heading:

'Bestobell — Controls, Energy, Aviation, Consumer Products — Worldwide'

Appendix II   Acquisition of Meterflow Limited

Anticipating the 1978 situation BML had acquired in 1977, through the main Bestobell group, a company called Meterflow Limited. It had been established as a private company in 1958 and was bought by EMI ten years later, but became almost moribund due to lack of incentive and interest by the owners. It was narrowly based as a manufacturer of turbine flowmeters which had a good name but not a very significant sales record. The company appealed to BML as a prospective growth business if more active management policies were introduced and it would help to complement the BML product range, particularly the Doppler flowmeter.

It was sited at Baldock and included a sheet metal shop which had been introduced by EMI to supply parts for its brain-scanner. This was now unnecessary but helped to reduce overheads by operating as a subcontracting business. Its machine shop suffered from obsolete tools, the test facilities were poor, and the control of stocks was unsatisfactory. There was a total staff of 85, a turnover of £316 000; and a net loss of £48 000 during 1977.

A general manager (Mr M.J. Scott) was appointed after take-over with directions to make the company profitable within a year. This proved quite impossible, but a realistic effort to give the company a healthy base from which to develop was undertaken. Mr Scott started by establishing a management team — something which did not previously exist. He was able to attract a young and enthusiastic marketing manager (Mr T. Cousins) who already had R & D experience at a large instrument company, who had a university background, and who had made flow-measurement his speciality. A new sales manager was recruited from Bestobell Mobrey and the existing production engineering manager was retained. This team was one that could make decisions when needed and provide the essential drive. It was backed by the experience of BML departments, but was in a somewhat anomalous position.

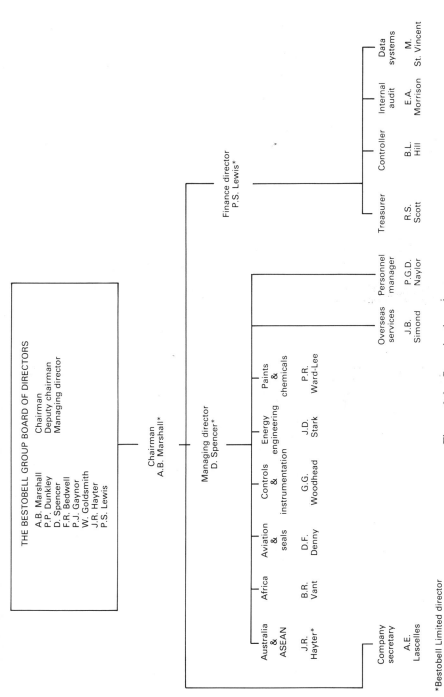

Figure 6.6 Organization chart

*Bestobell Limited director

Meterflow was nominally autonomous and had its own board, profits, and accounts. On the other hand, it was regarded as another department by BML and Mr Scott attending its executive meetings, made use of its sales force and personnel department, and consulted with its engineering and marketing staff. He had to follow BML procedure, for example, on job-assessments and other paperwork, which he felt was unnecessary and only led to large overheads. He was, nevertheless, glad to have a corporate plan prepared, largely by BML. This called for a sales increase to £1.225m in 1978 rising to £1.856m in 1983 at constant prices. This would then provide some 10 per cent of the BML net profit. The greatest increase was to come from new products and particularly by introducing microprocessor systems. Plans to do this by licensing the technology from the USA were proceeding with active encouragement from BML, which saw its own interest in this field also being served from the same source. Other expansions were to be aimed at the development of vortex meters and more sophisticated instruments for digital blending, custody transfer, and associated requirements.

All this growth was to be undertaken with comparatively little additional staff. It was hoped that the business and technical capabilities of the existing staff could be developed within the company rather than looking to outside recruitment, and that they would be able to accept the increased responsibilities. Direct labour should increase as more in-house manufacture was undertaken and this would improve the overheads. Finally, the sheet-metal shop would be hived off as a separate business.

Appendix III DS 77 Switch Justification

DS 77 Design

The current switch (G 350) is operated by the movement of a magnet attached to the end of a float arm close to a second magnet within the switch body. This magnet is pivoted so that movement of the first magnet past its face repels it. When the float magnet reaches an appropriate position, the switch magnet flicks across and its rotation opens or closes the contacts.

The DS 77 switch is similar in principle, but a small rod intervenes between the switch magnet and the switch. The rod is arranged to reduce the lever arm of the pivoting magnet and hence, for a given magnetic torque, the force that can be applied to the contacts is increased. The result is a more positive switch action. Many other advantages were gained and are itemized below.

Contact force	Considerably greater ·
Electrical rating	Higher
Vibration resistance	Much improved
Handling resistance	Increased shock resistance due to simplified insulation components

Terminations	Improved terminal access capabilities in some switch bodies
Assembly	Wider tolerances permissible due to push rod adjusters
Moving magnet assembly	No dust-generation to contaminate contacts
Electrical path	Only one set of contacts

Rationalization

By selecting three basic forms of the switch, which can all be accommodated within the main design dimensions, it is unnecessary to provide a host of specially modified variants. This is a great advantage for all departments of the company. Production will benefit from the greater quantities of common components. The large numbers will permit suppliers to offer parts produced by more efficient operations, leaving less machining to be done. Assembly will no longer be handicapped by multiple variations and improved flow techniques will be possible. Production control will have fewer different parts to order and process, and computer control will become a possibility. Sales will have an interchangeable and extended product range in which personnel will find it easier to be conversant with the total range and able to give a more informed service to customers. Engineering will be relieved of many production problems caused by the great number of variants and their time will be available for their rightful duties associated with new designs. Accounts and purchasing will also have to devote less time to switches and the latter will have the opportunity to seek multiple quotations where previously, due to the insulating material used, it was tied to one manufacturer.

Marketing

The replacement of the old switch with its limitations opens up new fields for market-exploitation. Not only will it be possible to offer technical operation superior to that of competitors in existing markets but it will be possible to attack new areas both at home and abroad. For example, it is desirable to obtain footholds with the overseas nuclear utilities before competitors, and to establish the reliability of Mobrey switches in the hostile vibration and temperature environments which distinguish modern marine and petrochemical fields. At home, the new switch should enable product managers to re-establish the company's superiority before serious competition can damage it further.

Finance

It has been assumed that the G 350 switch would continue to lose sales at the 5 per cent cumulative rate already existing, but that the introduction of the

DS 77 will halt the decline and enable sales to stabilize at the current level. Other assumptions have been made for the cost of units and the cost of tools. From these, it is deduced that the net result will be a small loss in 1980 and 1981, followed by an increasing rate of profit. By 1985, the incremental profit from holding sales level should reach £88 000 but, if the incremental sales were non-existent, there would be an adverse cash flow of £77 000. In addition, there are intangible benefits arising from rationalization, as already indicated. These will exist throughout the company in the form of reduced paperwork and increased efficiency and effectiveness.

Appendix IV   New Switch Concepts

The DS 77 switch was the result of one man's ideas on how to overcome the limitations on contact pressure existing in the original G 350 switch. It seemed to be such an advance over the G 350 that it was accepted and its development started without consideration of possible alternatives. Only some weeks later was the company's new value engineering section invited to review the requirements for the replacement switch. It operated under Mr B.R.A. Whiting, one of the Cranfield School of Management graduates recently engaged, who immediately arranged a meeting of the main project team members. His meeting was conducted in two phases:

(1) A speculation and evaluation session;
(2) An improvement and modification session.

Speculation and evaluation

The meeting was invited to describe the switch performance by two words, comprising a noun and a verb. The response was slow and little enthusiasm was displayed, possibly because the leader did not provide enough impetus or possibly because the whole idea was foreign to participants. Half a dozen ideas eventually emerged and Mr Whiting selected 'close contacts' on which to base the next stage, which was a brainstorming period on how to close contacts. This still lacked inspiration and produced much criticism. It might have been improved if the key words had been selected by Norman West or Peter Downing, who were most closely associated with the whole project. This would have avoided any danger of the 'Not Invented Here' syndrome rejecting new ideas. In addition, there was little scope for anything new as the concept of a conventional configuration within the existing body was already laid down.

Improvement and Modification

This period was devoted to highlighting problems and suggestions for improvements. It pointed out:

(1) Customer-dissatisfaction with wiring terminations to standard (S 01) body;
(2) Competitive units use microswitches as does the BML marine unit (S 196 body);
(3) Could the S 196 body be used with a new switch to provide more accessible terminals?
(4) Could a high-temperature microswitch be found to extend the capability of the marine switch to other applications?
(5) Could the composite flange on the S 01 not be applied to the S 196 body?

CHAPTER 7

Probe Engineering Company Limited (Case History No. 5)

One attraction of starting a career in manufacturing industries is the expectation that competition provides excitement and variety. Any risk of uncertainty is more than counterbalanced by the hope of fulfilment and advancement. Occasionally large companies do not live up to this image and appear bureaucratic; then the more adventuresome and resolute staff leave and launch their own companies. This case history deals with a situation of this kind.

Four engineers, all with young families, left a large organization to set up their own manufacturing business. Whether their decision sprang from a sense of frustration or from an unconscious desire for independence is perhaps immaterial. What is important is that the many difficulties that were inevitably encountered were surmounted by strength of character and hard work.

7.1 Introduction

The Probe Engineering Company is pleasantly situated on the outskirts of Cirencester and is housed in premises which were built in the autumn of 1978. The site covers an area of some 9000 ft^2 and comprises a factory, laboratory, drawing office, general office, and mechanical engineering workshop. The premises are light, pleasant, and airy. A plan of the ground floor and first floor is shown in Figures 7.1 and 7.2. The company is privately owned and four executive directors own equal shares of the £12 000 equity capital.

The company was started nine years ago and, in September 1980, employed sixty-seven people of whom eleven were concerned with development and design. During October many schedules were deferred or rescinded and immediate action had to be taken to contain costs, one consequence of which was a reduction in staff to forty-five. The organization chart is shown in Appendix 1. The company specializes in electronic products aimed to satisfy needs in agricultural, commercial, and municipal vehicles, automatic vending machines, and environmental monitor and control applications.

For a company less than ten years old its performance is distinguished by many unusual features. The four owners began by each contributing £50 and succeeded in achieving a positive cash-flow in the first and subsequent years. Their average age was then thirty-three. The rate of growth is good, especially when viewed against the UK's economic climate, and the turnover of £550 000 in 1978 increased to £702 000 in 1979. The 1980 turnover was expected to be £950 000, with an associated profit of £90 000, but such was the

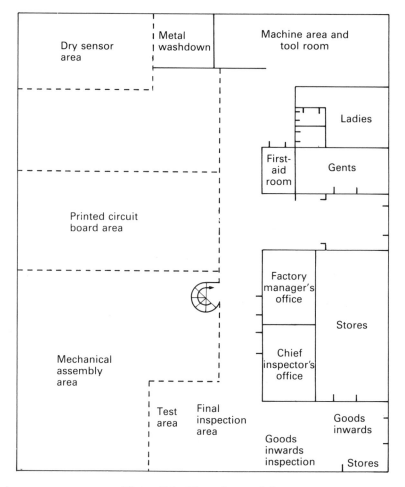

Figure 7.1   Plan of ground floor

severity of the downturn in business that only a nominal profit seems likely. The 1979 balance sheet is shown in Appendix II. The company is planning a growth of 5–10 per cent per annum in real terms, to be financed from internally generated funds. Despite an accumulated turnover of several million pounds, there has not been a single bad debt and, in explanation, the marketing director said: 'I try to understand the person with whom I am doing business and seek to recognize his motives.'

7.2   History

In the late 1960s five electrical engineers employed by an international group of electronic engineers were experiencing the nomadic life that seemed to be

134

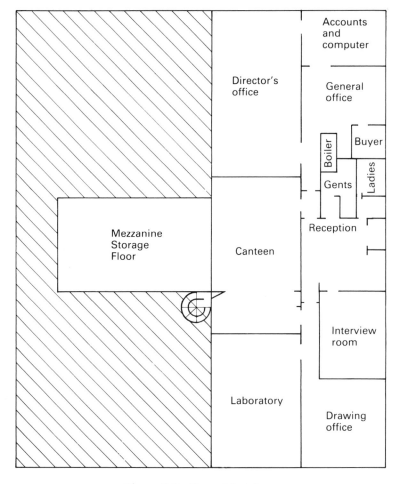

Figure 7.2   Plan of first floor

met with in many growth industries of that period. They were living in an hotel, waiting to be joined by their wives and young families. The men worked on microprocessor applications in a section termed 'assembled functions'. One of the team, Mr Roger West, was responsible for the marketing function of the section, the others were development engineers. Four of the five were newcomers to the group; the fifth, Mr Brian Green, had been transferred from another of its functions. The team was thrown into close association with each other and they had a great deal in common. The application of microelectronics was an exciting new field and the assembled functions team was keen to provide solutions for every expressed need. Not all proposals, however, fitted in with the group's market philosophy and the rejection of bright ideas was not uncommon.

One weekend in the autumn of 1969 the team met to discuss the rejection of their development solution for a Ministry of Defence problem. Their restless unsatisfied urge to translate their ideas into products boiled up and, following an animated discussion, a decision was taken to start their own company. Among the suggestions that were discussed, but rejected, was the possibility of manufacturing a product to meet the Ministry of Defence enquiry.

Mr Roger West's resolution to go it alone was so strong that he resigned from the group forthwith and accepted an appointment as a salesman with an international American corporation. His four friends continued in their employment for a short time, but worked on developing a new project in their evenings and throughout the weekends. One of the team, who had had previous experience in starting a business was not unnaturally less open-minded than his friends on how the new business should run and, following a friendly disagreement, left to set up again on his own.

The first market need which the team decided to meet was in the agricultural market segment and was to prove the forerunner of many later vehicle-borne devices. It began when a supplier of equipment to the agricultural industry, RDS Limited, a dabbler in electronics without much understanding, telephoned Mr West while still with the international group to discuss possible ways of improving the performance of combine harvesters. The concept of a corn-monitoring device seemed feasible and by the spring of 1970 a product had been designed (see Figure 7.3) and an order for 200 obtained. RDS Limited agreed to supply the components and progress was so rapid that the company was incorporated in July 1971.

Almost immediately after formation, the four men hired an empty chapel for £1.50 per week and each contributed their £50. In the following year, the small group moved to a yard and premises previously occupied by a builder, and took on one more staff (now the works manager) and three operatives. As production increased, a Portakabin was acquired. In the early days the commercial side was managed by Mr West and an accountant but, in 1977 Mr E.J.W. White joined the board as a non executive part-time director, in charge of finance. By 1978, Probe Engineering's growth was considerable and they were sufficiently prosperous to build and move into their present premises.

The above paragraphs might encourage the belief that starting a new company only needs a marketable idea and good fortune. Roger West's recollections, however, make clear the high reliance that must be placed on persistent effort and stamina. While working at the American corporation he not only had to make frequent visits to Europe but, when in England, needed to commute between Cirencester and Wembley. The M4 was not then built and a daily journey of five hours or more was entailed. He said:

...at one stage I decided to give up my ideas for our own company, since I was conscious of neglecting the family. I was persuaded not to do so by my

136

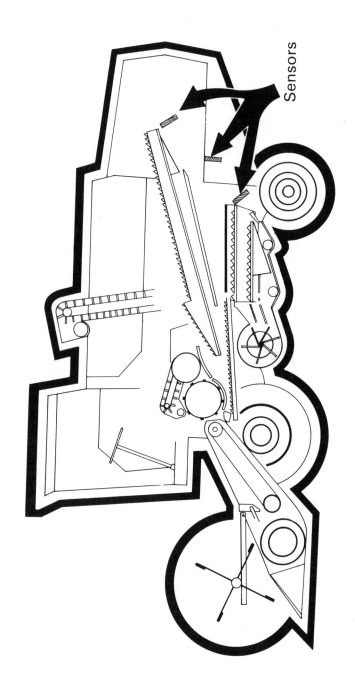

Sensors

Figure 7.3   Sectional view of RDS (agricultural) Mark 2 monitor

friends, who pointed out that, without my marketing experience, they could not continue and would have left their jobs for nothing.

Perhaps because of their early success none of the group ever entertained more than a momentary anxiety or doubt about the rightness of their decision to embark on their own enterprise, although 1981 is proving a very searching period.

With the past experience of Probe fresh in his mind, Mr West doubted whether he could ever again start another company. He was conscious of adopting a more elaborate and considered approach to any personal decision.

7.3 The Products

The range of products listed in Appendix IV is grouped according to whether a product is in production, is obsolete, or is obsolescent. For a comparatively small company, the number of possible products is large. The most important product, the corn-monitoring instrument, was designed to optimize the working of a combine harvester. This device, together with its associated sensor, accounts for half the company's turnover. It will be briefly described in order to illustrate the general nature of the vehicle-borne devices.

The problem is concerned with separating and collecting the corn when it and the chaff passes down and emerges from the discharge tube. The heavier corn falls into a bin and the lighter chaff is blown away, but the efficiency of the operation depends upon the forward speed of the tractor and the air-velocity down the tube. The tractor driver needs to know how to adjust one in relation to the other. The chosen method was to measure the difference in kinetic energy between the corn and chaff as they struck a measuring bar incorporating a transducer, to use a microprocessor to record the measured variables, and to translate them so that the driver knows which controls to operate. The device proved successful and is still being supplied as original equipment. It is an indication of the competence and thoroughness of the group that, after a few teething troubles the product proved completely reliable. This was to prove of crucial importance since had service problems arisen all over the country the reputation of the new company would have been irretrievably damaged.

The salt-monitoring device embraces a similar concept. The flow rate of salt from the discharge nozzle varies with moisture content and other factors, and the Probe instrument displays the forward vehicle-speed needed to maintain the optimum spread of salt on the road. The product is currently a victim of economies that are being made in local government expenditure.

The company policy is to design products for sale to customers who will take responsibility for marketing, selling, and servicing. Their taxi-meter can be cited as an extreme example of this wish to be solely concerned with design and production. The taxi-meter is a device which displays the fare as a single figure by computing supplementary charges. It can also be easily adjusted to take into account alterations to any of the rates, e.g. mileage, waiting time, luggage, and extra passengers. Since there was no obvious customer who would market the instrument, Probe took the initiative and formed a consortium with the Licensed Taxi Drivers Association (LTDA) and a major taxi proprietor. The taxi-meter development was, in many ways, a major departure for Probe, since considerable resources were used in obtaining a Public Carriage Office certificate and meeting British Standards Institute specifications. Four meters were submitted before approvals were obtained and Probe had to run a taxi many thousands of miles before the necessary reliability was achieved.

7.4 Funding

Although profit on the initial orders sufficed to convince the owners that the enterprise was viable, they had insufficient capital for expansion. Of several banks with which the possibility of funding was discussed, the Midland Bank offered the best terms and, in 1974, a loan was obtained of £18 000. Mr West also established an excellent relationship with the National Research and Development Corporation (NRDC) and derived help by using them to test ideas. He said: '...It is a good discipline to draw up business plans for NRDC to scrutinize.'

In 1979 NRDC was approached for help with the taxi-meter project, for which the development costs were estimated at £50 000. They agreed to pay 25 per cent. The remaining £37 500 was still too high to be funded internally and an application was made to the Product and Process Development Scheme (PPDS) of the Department of Industry, who agreed to meet 50 per cent of the outstanding sum. Mr West stated:

> ...the directors of Probe Engineering are committed to autonomy and regard their equity as sacrosanct and indeed, on these grounds, recently rejected an offer of funding from the National Enterprise Board.

Because of the company's past performance and satisfactory cash flow, funding is not a current problem and, indeed, their relations with the bank continue to be excellent and a further loan would be granted if required. For a young growing company Mr West believes that a redeemable equity would be preferable to a bank loan, since payment of creditors is deferred until the business makes a profit.

Commenting on the scale of finance needed, Mr West remarked that prior to the recession, Probe Engineering spent a substantial proportion of sales receipts on activities concerned with the future. They allocated 10 per cent to development and a further 5 per cent to capital expenditure.

7.5 Product Innovation

On the general question of product ideas, Mr West's view is that:

> I have a bag full of concepts and seek an interface with an individual or organization that will do the selling. In essence, our activities are restricted to our main expertise—electronic engineering design and manufacturing.

Up to now, contracts have arisen through friends and customers. In illustration of this, a clear *rapport* was established with a local vending company who introduced Mr West to a larger vending company from whom further contacts have led to prospects for several products in this growing field.

The source of new product ideas arises mainly during discussions between the company's customers or potential customers, and relies on Mr West's ability to hear and evaluate market needs. Suggestions for new products are embellished by the development staff, and the lay-staff also contribute. Compiling design specifications is a vital but difficult task; a potential customer is frequently bemused by the prospect of so many design options that he tends to vacillate and procrastinate.

In order to avoid uncontrolled diversification it has recently been decided to restrict development to products that will sell in three specialized market segments:

(1) Vehicle-borne electronics;
(2) Automatic vending;
(3) Agricultural electronics.

The concern is always to restrict designs to those products which can be economically made by batch-production, rather than to robotic mass-production methods, since the latter would attract competition from the large multinational companies. Products meeting this policy are usually sophisticated ones associated with a turnover of £50 000 to £200 000 per annum, and which can sell at a price that will return at least 20 per cent before tax and after component costs, work costs, and directors' remuneration. A product with any subsequent modifications should last from five to ten years with a maximum return halfway through its life.

The company avoids working too much at the sharp end of technology but nevertheless tries to incorporate the best and most up-to-date components from Japan and the USA. Care is taken in assessing the ubiquitous nature of advanced components, since the company is conscious that their designs must, in every respect, be credible to customers. Microprocessors are mostly incorporated during the design stage and, because the software is one of the imponderables of product development, Mr West believes that knowledge and brainpower are the two most important resources of the company. Mr West commented that '...the task of keeping up to date with advances in technology is not difficult'.

Information comes from component suppliers, mailing shots, calls by manufacturers' representatives, trade journals and, above all, listening to customers. Little or no use is made of professional societies, conferences, symposia, or scientific papers.

The basic laboratory equipment with which development work is carried out comprises a Motorola Exorciser system dedicated to the development and prototyping of systems using 6800 series microprocessors. It is readily extended to the series 14100/1100/1200 single-chip microprocessors and when a product requiring production on a sufficient scale is to be developed. There are additionally the usual array of electronic test equipment multi-meters, bridges, oscilloscopes, power supplies, oscillators, pulse generators, counter timers, etc. An important feature is an array of specially designed black boxes for fault-detection of both components and assembled products.

Technical competition or indeed competition of any kind has not yet been serious, and this is attributed to the company policy of designing products for discrete, specialized, market segments. A weakness which they had to guard against is allowing good ideas to languish—the syndrome which gave rise to Probe Engineering! It may be observed that though the market segments satisfy the stated product policy criteria they were arrived at largely through chance contacts. An alternative approach could be to examine all possible home and export market segments and consciously to look for applications that call for Probe Engineering's physical resources and skills. It is questionable whether a corporate strategy can be achieved through complete reliance upon a reactive and opportunistic approach.

The progression of an idea to a marketable product follows a fairly conventional route. Once an embryonic idea is obtained it is discussed by the four directors among whom 'the level of consensus is high and the level of disagreement is low'. Should a potential customer identify a need for which a radical solution is proposed, the matter is discussed by the board and, if a tentative decision is made to proceed, a business plan is drawn up. A new monthly 'product meeting' has been recently instituted and so far it has been concerned mainly with progressing developments, and only discusses customers' ideas if they are very tentative.

When a new project is launched, very close liaison is established with the possible customer and, if all goes according to expectations, a prototype product is ready for assessment in approximately two years. Formal planning techniques such as networks and resource-allocation are not used.

A consultant has been employed to help with the design of the taxi-meter with particular reference to the aesthetic appearance, but formalized techniques such as value analysis and value engineering are not practised. Mr West is planning to evaluate their appropriateness to the Probe method of working. The thinking behind this is that for Probe Engineering to grow and survive as an autonomous unit every opportunity must be taken to underpin their informal approach with suitable methodologies.

7.6 Subcontracting and Consultancy

Although Probe Engineering would prefer to use all its facilities on products for original equipment, subcontracting has developed into a significant subsidiary activity. The work has mostly come fortuitously. Mr West has an ambivalent attitude towards contract work. He remarked:

'With original equipment, we know the market place and are in control of our business but, with subcontractors, we even find it difficult to find out how long the business is likely to continue. Subcontractors also introduce a reject problem which we ascribe to poor communications between our engineers and customer engineers; a problem which we are still trying to solve.'

Again, Mr West said that contract work had not been essential during the past seven years, but agreed that it evened out cash flow, filled troughs in normal demand, and accounted for 20 per cent of the turnover. The work also had the potential of cross-fertilizing development ideas. The current sub-contracting business includes a radio-controlled traffic light system and a foetus probe.

Following a Ministry of Defence order for an electronic timer unit, the company gained the Ministry of Defence 0524 approval which conferred a seal of quality assurance and dependability. The company has also secured BSI approval. The professional standing of the company is also enhanced by their appointment as consultants in the Department of Industry scheme, MAPCON, whereby they advise companies wishing to instal microprocessor systems in either their manufacturing units or end-products. Though keen to do this work, Mr West strongly expresses the view that to separate a feasibility study from the design is never thoroughly satisfactory. Ideally, both should be done at the same time.

7.7. Operation

Control of company expenditure is effected through computerized monthly accounts and the accounts department function is shared between Mr Lees-Smith and Mr Brian Green. The factory is divided into activity areas and the cost of each is compared with monthly budget figures; the staff cost is calculated from time sheets which allocate activities on a half-day basis. The budget is based on the profit required in the forthcoming year and extrapolated product costs.

Invoices and stock records are similarly computerized, a refinement which is necessary since each order may deal with a hundred instruments, each containing a hundred components. Although variety has not so far been a major problem, it may be in the near future. The value of the parts stock is normally around £100 000 and would be much larger if it were not for the fact that products are made against orders for immediate delivery.

Staff have been obtained locally whenever possible and only in rare instances has there been resource to national advertising. Turnover among employees has always been negligible.

Mr West's great interest outside Probe Engineering is in education. He has recently overseen a work experience group from which, happily, three joined the company. He is a governor of the local technical college and is active in efforts to start a new university in Swindon. He finds that contacts in the educational world are of value when recruiting new staff.

Up to the present, only experienced people have been employed and there has been little need for staff training, although an exception has been made for four development staff who have been trained in microprocessors. In selecting staff for development work Mr Harbour said he looks for commitment to creative technology and absence of grandiose views on management. The staff on assembly are both male and female, but predominantly the latter. Marketing and selling is done by Mr West; there are neither salesmen nor sales engineers. Most sales are direct to manufacturers, but occasional use is made of agencies.

Exports are indirect through Probe's customers and amount to 70 per cent. Little or no direct exporting is undertaken because an early attempt was discouraging and because there are no staff suitably trained. However, negotiations are presently in train with a West German agency.

7.8 Staff Relationships

The origin of the company is reflected in the sharing of one office by the four executive directors. Another unusual feature is that the chairmanship of both the company and the board rotates among them with a four-year cycle.

Despite the good relationships which have so far been apparent, uncertainties and anxieties born of the current recession are now beginning to be disruptive. The company does not react as quickly as hitherto and, as mentioned above, it is becoming necessary to introduce more formality in their procedures. Even at the top, the traditional team spirit is being eroded. One of the four directors now rarely uses the communal office on the grounds that he needs more time to solve development problems at the bench. Mr West admits to a feeling of increasing isolation and believes it to stem from the fact that his marketing responsibilities cause him to decree actions on the grounds of expediency rather than consensus. Again, there has recently been a weakening of their interactions outside office hours.

It is a common experience that unless deterioration in staff relationships is quickly checked, organization failure becomes inevitable: it is important to examine Probe's reaction to developing difficulties.

An important factor in the early experience of the four directors was that each had his own development programme and controlled his own resources. An early exception was the recognition that Roger West should take responsibility for sales and marketing. As the company grew, Brian Green and Bernard Lees-Smith realized that the overlapping of each other's jobs was becoming intolerable and, in 1975, they sat down and, following a discussion, divided up their roles. They both saw that there was a natural split—Bernard Lees-Smith agreed to look after production and Brian Green accepted the buying function. Just as the four directors recognized that Roger West was a natural salesman, so there was complete agreement that John Harbour was the best development engineer. A further natural division of responsibilities was later carried out in the development section when Andrew Bird, the principal engineer, agreed to look after the microprocessor software, leaving John Harbour with the hardware.

However, the process of assigning responsibility does not seem to have gone far enough and, because of additional stresses, it is unlikely that the company can continue to respond to external challenges without further change. Too much overlap still exists and, for a small company, too many people still wish to be involved in development. One consequence is that there must be a risk of confusing customers who may be contacted by four or five different people from Probe. The confusion of responsibilities is being worsened by the board's reluctance to face up to conflict caused by the company's failure to maintain promises to customers, difficulties which to some degree are common in manufacturing industries but which are now highlighted by Probe's drop in their 1980 turnover. Development programmes are taking much longer than expected and development costs are getting out of hand, with the result that too much time is directed towards rectifying faltering situations. One serious consequence is that senior staff do not have the time to stand back and take a detached view of events.

Not unexpectedly, the main strains became centred on the marketing and development functions. The development engineer becomes unhappy about product-specifications and customer relations, while the marketing and sales section feels let down by broken promises.

While this type of disagreement is almost inevitable during a climate of business downturn, the atmosphere of consensus which has been a feature of the company since its foundation is having the unlikely effect of accentuating the difficulty. It seems that this issue has not been frankly discussed in management meetings, because no-one wishes to appear to be the first to rock the boat. In situations of this kind it is found that individual members are often unaware that other members of a team share the same fears and concerns and would feel a sense of relief on hearing that the problems are to be aired and resolved. A third party can often be helpful, since he will not be emotionally involved in the problems.

The beginning of this strained relationship seems fortunately restricted to a few and, indeed, a majority of the staff are taking action to form a social and sports club. An attitude survey among staff below board level showed communications to be excellent and many commented 'members of the board are always approachable'.

Because the Probe company has had long experience in an open style of management it will no doubt succeed in adjusting to its changing climate. Several possible options exist: the first is for the directors to resolve their difficulties through mutual, frank discussion. The second option is to avoid overlap by writing formal job descriptions that will closely define line and staff responsibilities. Authority should be invested in one individual for the crucial activities of drawing up product specifications, planning, and control of development. Advantage would also accrue by working within a stronger framework of general management systems and controls. A third option will be necessary should the directors not believe that they can meet customers' demands simply as a result of installing administration procedures. This step would involve the appointment of a managing director to set targets, measure progress, and generally accept responsibility for carrying out corporate decisions. The four founder-members could continue to rotate the office of chairman.

Towards the end of the study, the company was asked to specify its most important problems, and the following were shortlisted:

(1) The problem of stabilizing the specification of a new product; and
(2) The assignment of priorities to projects on a day-to-day basis,

both of which confirmed the findings of the brief attitude survey. It was therefore agreed to end this study by bringing together seven members of the staff and to hold a creative session with the object of problem-solving on (1) and (2). It was additionally hoped that this session would improve personal relationships between the directors and establish a procedure that would lessen the possibility of internal conflict. The meeting took the form of a synectics mini-workshop and the proceedings are described in Appendix V.

7.9 The Way Ahead

In 1980 a corporate strategy and a product strategy were drawn up by the board for the first time and three market segments were classified in detail, namely:

1. *Vehicle-borne electronics*
   1.1 Taxi-meter
   1.2 On-board weighing
   1.3 Taxi-meter derivative

2. *Automatic vending*
   2.1 BIWA (ref. a West German customer)

2.2 Rampack (a style of vending machine)

2.3 Hospital drug dispensing (vending machines concerned with control and security)

2.4 Brooke Bond (automatic vending machine)

2.5 Autonumis (local manufacture of vending machine)

3. *Agricultural electronics*

3.1 Forage tester (refined) for tobacco and other commodities

3.2 Shaft-speed monitoring

3.3 RDS products

The recognized strengths of the company were stated to be in its ability to design and produce vehicle-borne electronics that are used in a hostile environment of vibration and dust, and experience in developing and manufacturing electronics for automatic vending applications.

It was recognized that, to expand products and replace obsolescent ones, the rate at which the company required innovations was approximately one major device and two minor ones per year. Previous experience suggests that a product's life-cycle is five to ten years for a major product and two to five years for a minor product.

Decisions have yet to be made on the ideal balance between high-technology and low-technology microprocessor-based products and on whether a proportion of the output should comprise products for direct sale to the consumer.

Mr West was asked to project his thoughts to 1985 and to name the first six goals that came into his mind. They were:

(1) A continued requirement for new products, with many still in the pipeline.

(2) A return on capital greater than 20 per cent.

(3) Numbers employed up to 130.

(4) A turnover of at least £2m.

(5) A planned growth continuing, but not so big that the company was in danger of becoming impersonal.

(6) A high company morale, without any pressure for a union closed shop.

Appendix I Organization Chart

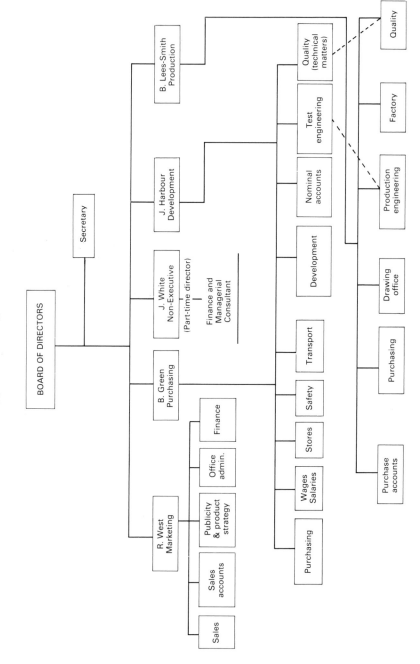

Appendix II Balance Sheet as at 31 December 1979

	1979		1978	
	£	£	£	£
CAPITAL EMPLOYED				
CURRENT ASSETS				
Stock in trade and work in progress	149 557		114 990	
Trade debtors and prepayments	221 850		167 013	
Building society investment	–		188	
Pension fund account	2		2	
Cash in hand	458		146	
Deposit account	2		1	
Taxation repayable	–	371 869	8 610	290 950
CURRENT LIABILITIES				
Trade and other creditors	95 807		44 591	
Bank overdraft	102 451		90 331	
Hire purchase creditors	22 234		9 094	
Directors' loan accounts	14 740		23 980	
Directors' remuneration outstanding	12 000	247 232	12 000	179 996
NET CURRENT ASSETS		124 637		110 954
FIXED ASSETS		194 137		168 717
TRADE INVESTMENTS		120		120
		£318 894		£279 791
REPRESENTING				
CAPITAL				
Authorized and issued				
12 000 ordinary shares of				
£1 each fully paid		12 000		12 000
MORTGAGE FACILITY		86 291		88 500
RESERVES AND UNDISTRIBUTED PROFIT		220 603		179 291
		£318 894		£279 791

Profit and Loss Account for the Year Ended 31 December 1979

	1979 £	1979 £	1978 £	1978 £
TURNOVER				
consisting of:				
Manufacturing sales	667 855		531 333	
Development contributions	34 164	702 019	19 436	550 769
TRADING PROFIT		58 188		50 023
after charging				
Depreciation of fixed assets	22 611		10 404	
Directors' emoluments	45 126		53 043	
Auditors' remuneration	2 000		1 700	
INTEREST		(25 761)		(12 808)
PROFIT ON SALE OF FIXED ASSETS		8 876		35
		41 303		37 250
TAXATION ON PROFIT FOR YEAR		(9)		(21 667)
		41 312		58 917
DIVIDENDS		nil		nil
PROFITS RETAINED		£41 312		£58 917

Statement of Source and Application of Funds for the Year Ended 31 December 1979

	1979 £	1979 £	1978 £	1978 £
SOURCE OF FUNDS				
Trading profit (before taxation)		58 188		50 023
Adjustments for items not involving movement of funds:				
Depreciation		22 611		10 404
TOTAL GENERATED FROM OPERATIONS		80 799		60 427
FUNDS FROM OTHER SOURCES				
Taxation refund		8 619		–
Mortgage facility		–		65 793
		89 418		126 220
APPLICATION OF FUNDS				
Trade investments	–		110	
Taxation paid	–		8 610	
Purchase of fixed assets less disposals	39 155		117 521	
Mortgage repayment	2 209		–	
Interest paid	25 761	67 125	12 808	139 049
		22 293		(12 829)
INCREASE IN WORKING CAPITAL				
Increase in stock	34 567		32 515	
Increase in debtors	54 837		57 813	
(Increase) in creditors	(64 356)		(11 364)	
(Increase)/decrease in directors' loans and unpaid remuneration	9 240	34 288	(4 740)	74 224
		(11 995)		(87 053)

Appendix III   An Analysis of 1978/9 Accounts

These notes offer some comments on the financial progress made by Probe Engineering as revealed in their 1979 annual accounts. There are many clues to the difficult year they were experiencing. Sales growth in the 1978/9 period (127.4 per cent) gives an appropriate idea of how, generally, we might expect other factors in the profit calculation to move. Total trading costs increased in line with sales (128.6 per cent) but interest charges rose 201.1 per cent, and retained profits were only 70.1 per cent of the previous year's. But creditors rose by 214.9 per cent, against an increase of only 132.8 per cent in debtors, indicating that current debt became more of a pressing problem during 1979. This is reinforced by the fact that total current liabilities rose 137.3 per cent during the year, while shareholders' funds grew only 121.6 per cent. In fact, current liabilities overtook shareholders' funds during the year. Considered as sources of finance funding the assets, current liabilities are the least stable of sources, requiring to be repaid and renewed many times a year, while shareholders' funds (share capital and reserves) are the most permanent 'bedrock' funds, never needing to be paid out while the firm lasts. At the start of the year, current liabilities were only 94.1 per cent of shareholders' funds but, at the year-end, they had reached 106.3 per cent. This means that shareholders were then risking less money in the business than the largely unsecured creditors. As stated above, the money owing to suppliers of goods and services more than doubled during the year. Faced with this, the directors reduced their emoluments for 1979 to 85.1 per cent of the 1978 figure. (Not many shopfloor workers would take a 15 per cent cut in wages!) It is fair to add that at the same time the directors withdrew loans of up to £9240, more than covering the £7917 shortfall in emoluments.

These considerations are more important than the traditional liquidity ratios, e.g. current assets/current liabilities. This ratio weakens from 1.62 to 1.50, which is not necessarily significant. It is interesting to note how little spare cash is kept available; the total of building society investment, pension fund account, cash in hand and deposit account, amounted to less than one day's sales both at the start and at the end of the year. The overdraft rose by £12 120, and might perhaps have risen more had not creditors been so patient! (We cannot, of course, be certain of this merely from the balance sheet. Perhaps a huge delivery of raw materials and components came in on 31 December.)

The most ominous single ratio is the 'times interest covered', the ratio of trading profit to interest charges. At the end of 1978 this stood uncomfortably low at 3.9 times (only £3.90 of trading profit for each £1.00 of interest) but had dropped to 2.3 times by the year-end. When we recall that a large company like J. Lyons and Company failed in 1978 with an interest coverage of only 1.2 we realize how urgent it is for Probe Engineering to reverse this trend.

These notes have focused on liquidity and solvency — 'cash flow' is the current term for it — because that is where Probe Engineering's problems lie. Profitability is not such an immediate problem, though the 1979 returns and margins did not beat inflation and the trends are down rather than up. Return on capital employed (trading profit to net assets) dropped from 10.9 per cent to 10.3 per cent. Capital turnover in pure money terms showed a slight quickening from 1.20 to 1.24 times, but failed to compensate for the fact that average margins were squeezed from 9.1 per cent to 8.3 per cent.

To sum up, Probe Engineering, for all its lively innovation and committed owner/management, spent 1979 gradually falling a victim to inflation and cash-flow shortage on both current and long term. The deepening of the recession during 1980 will not have made things any easier.

Appendix IV   Product Range

Presently in Production

MOD digital timers MK4 TC
Automatic cluster removal control box
Foetal monitor
Moisture-monitor sensors
Wheel sensors
MK3 combine monitor meter unit
MK4 combine monitor meter unit
Hectacremeter
Massey Ferguson shaft speed five-channel and four-channel
John Deer shaft speed five-channel and four-channel
Crop store temperature sensors
Dry sensor
Speed area meter
Speed area flow meter
Radio-controlled traffic lights
Taxi-meter

Not Presently Made, but Full Documentation Available

Flotation test jar equipment
Waste water flow meter
Laverda shaft speed five-channel
Distance recorder
Temperature probe

No Longer Made and Would Require Re-engineering

MK1 combine monitor meter unit
MK2 combine monitor meter unit
MK5 combine monitor meter unit
Wet sensor
Revtel bearing speed recorder
Salt alarm
Forage meter
Rally car split-time clock
Pad light

Appendix V   Report on a Synectics Mini-workshop, 1 April 1981

For a description of the synectics process see Chapter 3, page 46.

Session 1

Problem

To stabilize the specification of the product.

Springboards

1 Divide into fixed and variable components of specification.
2 Define specific design authority.
3 Get people to commit in writing what they really want.
4 Arrange a time every day when you will talk to them and they to you.
5 Update specification at regular intervals.
6 Delegate someone to actually write the specification.
7 Get a more direct line of contact with the final customer.
8 I wish our customer was as open to us as he would like his customers to be to him.
9 Have someone not directly involved in development sit in on the meetings.
10 Improve internal channels of communication.
11 How to relay and get feedback.
12 How to make sure customer or design authority has full knowledge of alternative products.
13 How not to worry about customer.
14 How to make a product to our own specification and then sell it.
15 Associate each change with a penalty.
16 Design the product with a more variable specification at the outset.
17 How to make customer more aware earlier of possible trade-offs.
18 Give them away free and make people pay when satisfied.
19 Shoot customer, and take over business, sell directly.
20 Go to our customer and threaten him to take product or be shot.
21 Have instant information block transfer — kids' stereo viewers.
22 Tell customer to take it or leave it.
23 Threaten customer to charge on cost plans basis.
24 Have a meeting and tell customer specification will not change — they must buy whether it works or not.
25 Sell something with a built-in specification (vegetables).
26 Get specification written up by 15-year-old.
27 Have a shop where we sell our products.

28 Have expunged from public memory the man who said microprocessors can do everything.
29 Buy microprocessor from that guy.
30 Sell the advantage of a restricted microprocessor.
31 Expand time-scales to fulfil specification by borrowing.
32 Buy a microprocessor to do specification we want and call it something different.
33 Make universal product.

	PRIORITY CHOICES	
ROGER	3	11
JOHN	1	33
DAVE	3	
TOM	27	26
KAREN	3	33
MARGARET	2	10
MIKE	3	27
BUNNY	6	7
BRIAN	6	1

Session 2

Ideas on No. 3

1 Publicity — make people aware of what Probe is trying to do.
2 Compose a form they get filled in by customer.
3 Have bigger sales force to knock on doors.
4 Give a 2-day customer-familiarization course.
5 Market research into what people want and you can provide.
6 Any proposal that came from checklist should be established design parameters.
7 From checklist, decide where risk lay on continuum of uncertainty and benefit.
8 Target date for reporting back on preliminary investigation.

Action — Roger to initiate

• To provide a checklist of a generic target specification allied to a specific range of products — experiment with it. Development department would need to highlight constraints of checklist — also procurement information.

Need

• Better ways of stopping risk from being risk.

Idea

- Stick to things you know we can do.

Concern

- How to ensure long-term survival when focusing on non-development products.

Possible Change of Policy (John)

- Look for unpublished information, researched by examination of competitor's product.

Idea

- Try to agree on what risk is.

Examples of Risk in Music and Sea

- Drop transistor in sea, it won't work
- Drowning
- Risk of pop record being a success
- Risk of playing music loan — drifting
- Wrong note
- Submergence of New Holland by sea
- Being caught in a storm
- Eaten by a shark
- Unfinished symphony
- Dies and becoming famous afterwards
- Drowning on day published
- Not arriving
- Seasick

Absurd Solution

- Better rehearsals
- Swim in safe waters
- Take a seasick remedy
- Edit recording tapes very well
- Look at weather chart
- Watch out for chewed-up sharks
- Go by rail

Ideas for Reducing Risk

1 Find customers who are not so clever/more gullible
2 Find cleverer customers
3 Insurance policy for failed products
4 Start on more than one product at a time
5 Stick to what you know

Session 3

Insurance

- Identify and exploit spin-off ideas
- Get somebody else to pay for feasibility of development
- Take on a number of non-development products
- Get development work done cheaply, e.g. as post-graduate student exercise
- Somebody else to do the development — act as middle-man
- Undertake a tremendous number of development projects
- Hedge the bet
- Off-load the high-risk ones — re-insure
- Get the customer to do the development work and we assemble
- Get customer to put up money to share the risk
- Approach an insurance company
- Try to quantify the risk ourselves in terms of odds: Return vs risk ratio for each project
- Measure attractiveness of the project
- Have several people developing the idea independently and then bring them together
- Insure with Lloyds — rainfall — analysis of past performance

Proposal

- Do not tackle any job unless tooling and development is funded

---

1 How to think of an idea early enough
2 How to convince some independent money-providing institution to finance deal, so as to prevent market opportunity being restricted to that customer
3 How to process procedure quickly
4 How to make provision for high-risk high-return projects

Ideas on 1

- Send out more reps
- Penalties for exceeding development period
- Consult an astrologer
- Have area on check sheet for people to mention ideas for the future
- Get ideas from people in company, with incentives, about product opportunities
- Study of market trends
- Change market desire ourselves
- Create the product

Idea

- Divide projects between those for which customer pays for development — large number/low risk and those for which company pays — small number/high risk

DIVISION OF DEVELOPMENT CAPACITY		
	Customer funded	Company funded
MIKE	70	30
MARGARET	60	40
KAREN	70	30
TOM	80	20
DAVE	75	25
JOHN	50	50
ROGER	90	10
BUNNY	70	30
BRIAN	60	40
	40	60 if own product

Action — Roger

- Have a business plan for every development project
- Devise self-funded projects in same way as we would devise a project to be funded by someone else

Session 4

Summary of New Actions

JOHN

- Development department has agreed to contribute to a checklist which would isolate features which were potentially risky and make customer realize what he is looking for — get us early information

ROGER
- Prepare business plan for various opportunities we see, which appear worthy of business plan
- Devise customer checklist

TOM
- Course for customers
- Tom to badger John

BUNNY
- Do market research

BRIAN
- Set up a scheme whereby employees' ideas on future products can be evaluated

CHAPTER 8

A Manufacturer of Components for the Automobile and Process Industries (Case History No. 6)

To accept the challenge of turning round an ailing company demands vision, ability, and great energy. When, moreover, an initial examination shows a rapidly deteriorating financial future to be compounded by poor products and bad labour relations, the task is one of unusual difficulty. In addition to these virtues, emphasis must be placed upon qualities of courage and leadership. Such a situation formed the subject of the following study. The main problems are brought into sharp focus because the observations that are described occurred during a period of severe economic recession.

8.1 Introduction

In 1978 this company employed 1300 people and its business primarily concerned heavy presswork, the manufacture of components for cars, commercial vehicles, and the process industry. In the past much of this had been centred solely on the automobile industry, but the firm was not geared to mass-production methods and a great deal of the private car market was lost to the industry itself. In more recent years, effort has been directed towards heavier vehicles and process plant. The total turnover was about £18m per annum and the net result, after tax, showed a slight improvement during 1975–7 from a loss of approximately £1.0m to a profit of £0.5m.

The firm is a subsidiary of a large group which covers a wide range of engineering components and has a turnover of more than £320m. Its main group contact is, however, with the central research—Ashworth House. As this is not far distant, a reasonably close association has been maintained. The group has been concerned with the future of its subsidiary and has appeared uncertain of how to proceed. In 1976 it offered to sell the company to a competitor, but this was turned down. It later made a bid to buy the competitor, but this also was rejected. Not surprisingly, morale weakened but is slowly recovering in the light of the group board decision to continue its backing.

8.2 Company Background

The company was started in the late 1880s as a private company manufacturing pressed metal parts for bicycles and, later, parts for automobile components. A rapid expansion occurred after World War II and in 1951 it was sold to the present owners. At that stage, profitability was some £¾m on a turnover of £10m. Although the new owners put in a great deal of money and doubled the production floor space, the financial performance declined steadily until the late 1960s, when losses were running at £1m per annum. This appears to have been caused by a lack of appreciation of the seriousness of the competition and the actions necessary to maintain an adequate share of the market. The outlook appeared to be very short term, planning was weak, overheads poorly controlled, and labour relations bad. The latter was partly attributable to the widespread poor industrial relations of the region and use of old machines and poor handling equipment. Stoppages of five to six weeks at a time were constantly threatening the company business.

The short-term policy is illustrated by the attitude adopted to new projects. A suitable material was proposed by Ashworth House which, although it conferred intrinsic advantages on the product, had associated processing problems. Research carried out between 1965 and 1967 provided a possibly promising solution and progress was such that, in the late 1960s, a small-scale production unit was set up and high-value components were supplied to a number of small-production specialized vehicles. Unfortunately, by 1970 the company was accumulating losses of £1m per annum, and a new managing director was appointed. His contributions to the company's future were significant. He began to raise morale, he inspired hope for the future, and appointed a number of able staff. Development programmes on current production were initiated and an important decision was made to manufacture a critical bought-in component. A decision was also made to achieve large-scale production of their 'substitute' raw material product. This, however, was not successful, partly because the scale of investment needed to achieve a cost target could not be risked at a time when retrenchment seemed unavoidable and partly because there was a temporary reversal in the price of the two raw materials. The situation during this period remained extremely serious, with few products to sell, prices too high, a declining labour force, and a continuing awareness of the difficult economic outlook. In 1975, another change of managing director was made and a fresh attempt made to revive the company.

8.3 Recent Policy

The new managing director, Mr T.D. Wilkins, inherited an organization with many problems arising from the failure to persist with the 'substitute' raw material product. He was an engineer by training, young, enthusiastic, and believing in 'management' as the key to success. He was receptive to any ideas for improvement and it was on this count that he became keen to encourage the CEI study. His view was that 'a disinterested, independent observer' could provide valuable help.

His first actions in the company were, however, to continue a policy of retrenchment and concentration on such markets and products as were still viable. He terminated the 'substitute' project and suspended all developments or other innovative proposals. A new company structure was introduced and, whereas it had previously been organized on a functional basis, it was now arranged into manufacturing divisions based on autonomous profit centres, each with its own accounting and marketing functions. These were to be supplemented by independent administrative sections supplying a service throughout the company. At the same time, he encouraged 'an honest, frank and open manner' at all levels. He introduced a consensus approach which permitted interaction between departments, and developed personal co-operation amongst all staff. He also stimulated shopfloor motivation through this general management attitude by revealing financial information about the company and by introducing profit-sharing schemes.

By 1977, these changes helped to turn the earlier losses into a small profit and the managing director was able to adopt a more optimistic outlook. For the first time, a longer-term corporate plan was prepared. This was largely a general policy statement and laid down only broad objectives such as 'to remain in profit and to avoid consuming cash', 'to protect the traditional business while expanding the market base and developing the product range', and 'to maximize sales of capital plan equipment in the process engineering sector'.

It did, however, give slightly more guidance on the individual divisions' objectives. For example, it separated the 'A' division (see below) aims into automotive, industrial, and capital plant headings, and laid down a policy of increasing sales by competitive pricing, by increased field sales force, and by improved delivery times. It also hinted at some degree of development and innovation by suggesting how the markets might be broadened.

There was no further elaboration on the execution of the corporate plan, but a five-year forecast was prepared on the assumption that the general directive would be followed successfully. It showed sales increasing by 75 per cent over the period, with profit before interest charges rising from £0.42m in 1978 to £4.3m in 1982. There was, however, a cautionary note indicating that 'based upon historical achievement, it must be seen as considerably speculative since a significant proportion is dependent on sales volume increases which derive from conquest business'. With such indicators, there was a clear signal to the manufacturing divisions to pursue vigorous expansion policies but, surprisingly, communication with the lower management levels seems to have been inadequate and many middle managers were unaware of the existence of a corporate plan and its implications.

Another development being planned was the introduction of the added-value concept to the company accounts, but this needed preliminary education of shopfloor workers. It was hoped that, by lucid explanation, their motivation would be increased and they would appreciate that the higher the added-value, the higher would be their wages.

8.4 Company Structure

The new company organization consisted of three manufacturing divisions—'A', 'B', and 'C' (see Appendix I). Each of these had executive managers controlling particular functions such as engineering, manufacturing, marketing, and accounting. Although not all of them were of equal status—several were directors of the company—they were relatively independent and all were directly responsible to the managing director. They operated through section managers, most of whom were comparatively young and who had joined the firm relatively recently.

The arrangement involved a certain amount of mixed responsibility as, for example, in the dual role of Mr R. Corbin who was general manager (manufacturing) of both 'A' and 'C' divisions, and in the relationship of divisional accountants and the company financial director. This lack of clearly defined limits of responsibility descended to the middle management. Nevertheless, the whole arrangement seemed to work without clashes of interest and with a degree of co-operation which was remarkable.

By the end of 1977, the company turnover was about £18m, with 'A' and 'B' divisions each responsible for some £8.4m, and 'C' division for the remainder. The latter designed and manufactured systems for commercial premises. It had growing sales until the energy crisis of 1973 when the market collapsed, and has since shown little recovery. The divisions' difficulties have been increased by a somewhat uncompetitive position. Although its products are of high quality, they are much heavier than those of any rivals, and are very highly priced. Furthermore, the products are manufactured under license and have not been able to produce a profit. Their only value is as a contribution to the company overheads and the future of 'B' division is a matter of serious concern.

'B' division is a supplier of press parts to the automotive industry. It is geared to comparatively low-volume, high-price output and does not compete with the larger concerns. Its main advantage is its flexibility in manufacturing small batches and its main limitation is that it has no forging or turning facilities but has to rely on fabricated constructions. It does not provide a design service, but regards itself only as a subcontracting business which gives a high return on assets. Over 80 per cent of its business is with two customers.

'A' division aims at three major markets, all of approximately the same size. The first sector covers requirements for passenger cars, commercial vehicles, and for off-the-highway vehicles. Although the private car business has been largely lost, the remainder still represents a turnover of more than £2½m per annum. The second is in the process industry, and the third is for large-scale capital plant. The turnover in each of these sectors is of the order of £3m per annum.

The large plants were mainly for developing countries in the Middle East and South America and were of the one-off variety. The result was that they involved greater technical staff time in dealing with individual design problems. The staff required was thus much larger and with generally higher qualifications than in the other divisions. It numbered some fifty-seven, mostly graduates, compared with twenty, none of whom were graduates, in presswork. Until recently, it had had no innovative function and its main effort was confined to short-term developments.

8.5 'A' Division

When it was recognized that it was necessary to introduce new ideas into the business if it were to survive and expand, 'A' division was acknowledged as the obvious place to develop such a policy. It has had a chequered progress in the past on new projects, but it had a range of products which could well lend themselves to innovative actions. Its engineering staff was strong in both theoretical and practical background and, although it had been discouraged from any enterprise by the previous short-term policies, the new directives gave some stimulation and encouragement for it to demonstrate its ability. With this in mind, it was agreed that the case study should concentrate on this division.

As indicated in section 8.4, the division was organized into four executive branches. Manufacturing, marketing, and accounting played a relatively subsidiary role in the innovative process and the main burden fell upon the engineering branch. This branch was divided (see Appendix II) into a product development group, a process engineering group, an applications group, and drawing office, planning, and quality control. It was headed by Mr A. Jenson, a powerful, red-headed Scot, who was ambitious, autocratic, very energetic, and intelligent. He was an engineer and a director of the company. He had previously been employed in industry and had also had experience as a consultant. This made him suspicious of studies which might be regarded as any form of disguised consultancy. He believed that such studies, if left in the hands of younger graduates, would only involve a waste of his time in educating them. However, he welcomed the CEI study, provided that the utmost tact was used in the investigation. This was because he felt that his staff was already suffering a trauma caused by his own drastic reorganization, and 'a period of undisturbed progress was essential'.

After early discussions, he had little direct contact with the study and, in fact, left the company in August 1978. The main liaison was with Dr C.D. Powell, who was head of the product development group. He was a tall, quiet, refined man, with a cautious attitude to new projects, often stressing the difficulties in progressing new ideas. He had previously been on the staff at Ashworth House and was well versed in research methods and the technical problems of 'A' division's products. His group was responsible for the early stages of all new ideas and projects. These could be initiated from many sources: for example, from Ashworth House research, from the applications group, from the examination of competitors' products, or from their own outside observations. In connection with the last, Dr Powell and

Mr Jenson had recently toured the USA examining the market and looking for possible license arrangements. A major study of the European scene had also been completed and a visit to Australia made.

The second group within the engineering branch was process engineering under Mr M. Dutton. He was obviously highly competent with a great deal of valuable expertise, but had only been at the company eight months and seemed to be a little uncertain on his precise functions and authority, probably because of the interactive and overlapping system. His group was responsible for some aspects of design, for some engineering development, and for providing improved manufacturing facilities. It might be regarded as the practical arm to complement the product development theoretical arm. The latter had the main links with the research work carried out at Ashworth House, and the former the links with the general production and manufacturing department under Mr R. Corbin, who also controlled 'C' division's operations, but not those of 'B' division. He was a director of equal status with Mr Jenson.

The third group with the branch was applications. It was responsible for receiving enquiries, seeking new markets and planning products, studying technical aspects and reliability, and liaising with the sales department. It was headed by Mr L.E. Bales, another fairly dynamic personality, who had had some eight years R & D experience within the group. He could require Dr Powell to undertake exploratory work of up to five man-days on possible projects, and this was usually carried out with a minimum of formality and with maximum co-operation. Any work requiring long duration was first discussed with Mr Jenson, as were all questions of priority.

The three groups had an excellent degree of co-operation and it was noteworthy that there appeared to be a complete lack of friction between product engineering and process engineering. This was fortunate, as their work was so interrelated that any arbitrary division could have led to serious hold-ups. Much of this good situation was undoubtedly due to the personal relations between Dr Powell and Mr M. Dutton. The latter left the company at the end of 1979 and was succeeded by his deputy, Mr T. Pratten. At the same time, Mr Bales (who had replaced Mr Jenson as head of engineering in 1978) introduced a move towards a more rational arrangement by changing the name of 'process engineering' to 'production engineering'.

The remaining two groups had fairly routine duties and had little influence on new projects or innovations. The drawing office employed about seventeen people and dealt with day-to-day aspects of work such as providing cheap but efficient jigs and tools for small batches, rectifications, and other routine demands. Its head was Mr S. Hibbert, who had been with the company for many years. He was extremely loyal and his long experience made him a sound choice for a fairly straightforward department. Similarly, the quality control group had no direct involvement with design but could influence drawing-release procedure. It tried to maintain a balance between quality control and product performance and cost. It was headed by Mr L.F.

Marson, who had taken charge about one year earlier after ten years in the firm in varied areas—technical sales, design, application engineering, etc. He had raised the status of his group, particularly by achieving a certificate of approval from the Ministry of Defence—regarded as a 'blue riband' in quality control. This gave the firm a prestige with potential customers, especially overseas, where quality was a prime requirement.

8.6 Ashworth House

Although Ashworth House was not part of the firm it had a very close association with it, particularly in its links with 'A' division's activities. It was the headquarters of the group's centralized research activities and undertook work on behalf of the various operating companies within the group as well as initiating more general developments. It had a number of specialist departments covering many disciplines, e.g. physics, mathematics, electronics, chemistry, and engineering. Its basic income came from a levy on each company, for which services were provided free up to the levy limit. Above this, companies could contract for additional work or seek help elsewhere. This company, for example, contributed £100 000 in levy and additional contracts during 1977, but also used NEL and Harwell as sources of research information. Its contacts with Ashworth House, however, were generally good and could be attributed to two main factors: the first was the proximity of the two establishments and the second was the interchange of staff that occurred between them. This applied particularly to 'A' division. Both Dr Powell and his deputy, Mr M.N. Watkinson, had had long experience at Ashworth House, as also had Mr Bales. On the other hand, Dr S.T. Gruber, head of the Ashworth House department concerned with 'A' division's technologies, and his deputy, Mr W. Donnington, were well versed in the products of competitors. The result should have been an almost copybook communication condition but it suffered because of the exceptional common background of Dr Powell and Dr Gruber. The former's specialized knowledge was apt to result in some unnecessary detailed discussions and the latter wanted to isolate his department to avoid infection from a too-parochial attitude by the company. As a result, although work was allocated to Ashworth House by the company there was a tendency for Dr Powell to inject his ideas or to do parallel work without reference to Dr Gruber. This induced some reaction which was, fortunately, balanced by the ability of Mr Donnington to maintain a smooth exchange.

This was later confirmed when, in early 1979, the group decided to decentralize the Ashworth House research work and more than sixty staff were transferred elsewhere or made redundant. The company's activities were stopped or returned to Dr Powell, who was permitted to recruit some of the redundant staff. In particular he took on Mr Donnington and so gained 100 per cent of his services, whereas previously only 40 per cent had been available. Not surprisingly, Dr Powell reported that the move had been unbelievably successful. He also acquired four engineers and some of the specialized test equipment.

8.7 Development and Innovation

Before 1977 the company had been restricted to a limited development programme aimed mainly at its existing market needs but, as already noted, this was then expanded with the object of achieving major future product-improvements. The great part of all the R & D work was directed by the product development section of 'A' division and its tasks ranged from 'survival efforts' through short-term product developments to the longer-term innovation projects. There appeared to be no clear-cut divisions between these tasks and, as with the overlap in organization, so there was a similar lack of definition on development categories.

In general, any work with research implications would be allocated to Ashworth House and the company team took on the more limited-duration items. The former were given contracts for fixed sums, varying between £500 and £25 000, for specified jobs and work continued until the allocation was exhausted or renewed. Most projects were of medium-term duration, but work was controlled by a series of limited short-duration targets. These were usually of one month only and were reviewed by committee meetings at the working level. Progress was also reviewed at three-monthly intervals at a more senior level and, finally, at six-monthly intervals when company executives discussed policy. The short-term objectives seemed to replace any network schedules and inhibited the innovative aspects. In most cases, work proceeded along technical development lines rather than as scientific research.

In 1977 there were about twelve identified product-development areas which were regarded as important enough to be allocated separate headings for accounting. Of these, three might be designated as innovative and long term, six were development or short-term items, and the remainder routine or continuing work for survival. The last included, for example, the evaluation of competitors' products (which took more than 25 per cent of available resources) and studies of product-reliability.

The short-term developments were intended to be completed in a few months but often extended longer. Of those itemized in January 1977 only three were completed within twelve months. Many new ones were added in 1978 to replace the completed ones, and it was worth noting that these seemed to have wider implications, such as value-engineering considerations, than the earlier ones, which were confined to narrower objectives. The longer-term work was concerned with major developments of their basic products and here there was much more creative thinking. This is the subject of the remainder of this case study.

8.8 'B' Projects

Due to the years of neglect many of the company's 'B' products were inferior to those of their European and UK competitors. The position was stated clearly by Dr Powell in a proposal to 'A' division's engineering manager, Mr Jenson, in April 1977:

> The company can currently offer no designs whose basic principles are capable of competing with those available to our competitors.

He pointed out that one way to deal with this situation was to manufacture a 'Chinese' copy of the best available competitor design, but that this would not give the company any competitive advantage. What was required was a 'next-generation' product with a marked superiority over the best currently available.

To this end Dr Powell suggested that earlier work carried out at Ashworth House should be re-examined and, hopefully, extended. The concept was to establish theoretical design principles that could be applied to a number of different and important products. The suggestion was eagerly accepted and steps taken to develop it by commissioning Ashworth House to carry out the theoretical studies. This study was, at a later date, code-named 'Advance'.

Two further ideas for improving products emerged from the product development group and two of these were followed up in 1978. The first was to modify the so-called 'BU' product by using a novel design and a programme to demonstrate its feasibility was initiated.

The third proposal involved a much more critical decision since it concerned a development whose success was imperative if the company were to retain any credibility in their main market. The decision was to anticipate some of the results from the advanced product in the hope that there would be sufficient creative anticipation to enable this competition to be leap-frogged. It was termed the 'Sink or Swim' project.

These three projects were all innovative in character but covered different degrees of company involvement. The first could be regarded as part of the longer-term policy to supply new research development knowledge for improving future products, the second as a project assessment which was still not advanced enough to justify all-out support, and the third as a major engineering project which could involve all the company's facilities. The progress on all three is discussed in the following sections.

8.9 'Advance' Studies

The theoretical design studies concerned the optimization of a system whose two design plants were in conflict with each other. The proposed solution was to introduce four nominal interacting design variables and see whether they could be specified independently.

To investigate the idea it was necessary to establish the performance characteristics of the variables and the problems of manufacturing. This was contracted to Ashworth House in April 1977, with target dates for manufacture of a first test-piece by October 1977, and a performance-assessment of all the variables by January 1978. To assess their ideas Dr Powell estimated that thirty-six samples would cover the range of four variables and that each sample would require one man-week for manufacture and test. Added to this would be the cost of tools and staff time in analyzing and reporting the results. Altogether the finance required should be £9000. Ashworth House took a much less optimistic view of both time and cost involved and estimated that at least £18000 would be required. As a first exploratory step, it suggested spending only £1500 in developing a tool for one sample and testing this. In the event, even this proved to cost twice the estimate but it gave a promising test result.

Much of the trouble lay in the production area. There was one particularly difficult operation which concerned the manipulation and forming of materials. No easy solution was forthcoming, but Ashworth House had had some experience of one technique and recommended its aceptance. Dr Powell's team, on the other hand, thought a more original approach should be undertaken that would be more suitable for maximum production. A compromise was agreed. In order to obtain performance results early, the Ashworth House system was adopted for a limited number of samples (nine), while work on the more innovative approach would be tackled jointly by the research centre and the company. All this set back the original target dates by four to six months and raised the finance required to nearly £14000.

At the end of twelve months, however, there was still insufficient performance data and the manufacturing problems had not been overcome. Only six samples suitable for test had been made by the first technique and one by the second. The latter showed some promise but had the anticipated production snags. Ashworth House, whose main interest was in producing the theoretical data for design parameters, believed that future manufacturing needs were Dr Powell's responsibility and he persisted with his early idea.

By the end of September 1978 the project was eight months behind the original time for completion, had cost £7000 more than expected, and had still not produced a clear outcome. A report was prepared on the tests completed with an analysis of the data. This concluded that, in spite of the wide variations in performance of nominally identical samples, there was enough advantage to justify further development work. The company agreed to pursue the project, adopting Dr Powell's concept which, if successful, would permit continuous production. A further budget of £12000 was allocated for the 1978/9 financial year and a new target set for a full assessment of design potential and manufacturing capabilities by September 1979.

The work seemed to lose some impetus at this point and progress became a slow trial-and-error procedure. For three months various ideas were tested with little success. In early 1979 there was a serious interruption caused by a policy change at group level. Ashworth House was directed to divest itself of many of the activities undertaken on behalf of the group. As a result, the company found itself responsible for all further work on 'Advance' but was able to recruit some of the redundant staff from Ashworth House.

After reappraising the situation, Dr Powell proposed continuing the project but with a less ambitious programme. The project would be attacked in an evolutionary way, starting with examination of the most simple form. The work was allocated to Mr Donnington, the senior engineer who was transferred from Ashworth House. This proved to be a very successful arrangement, as co-operation improved greatly and, over a period, results on both performance and manufacture also improved. By the end of 1979 Dr Powell was able to express the opinion that 'the mathematical background and procedures were sufficiently sound for the design of new products for novel applications'. The work was allowed to continue as an exercise which was expected to produce important spin-offs for all 'B' projects.

8.10   The 'BU' Project

The 'BU' project involved an engineering solution whereby performance, strength, and cost of a product could be achieved by applying one design modification that was based on a new theoretical concept. Once again, however, the idea, although attractive, was associated with difficult engineering problems. No special interest was taken in the idea by the applications group, but the production development group decided to investigate its possibilities and Mr Watkinson, deputy to Dr Powell, was given responsibility for this. He undertook a general theoretical analysis of design parameters and, in early 1977, was able to lay down the design requirements which would have to be met if the existing product was to be replaced.

To establish the validity of the theoretical work the next step would be to carry out tests with samples which could be subjected to operating conditions. Product development decided to continue the investigation through this stage still relying largely on its own resources and a programme was prepared. This entailed the provision of tools and the manufacture of nine small samples which could be tested both thermally and struturally. It was believed that the tests could be commenced by October 1977, and completed by the end of the year. In fact, manufacturing problems delayed the testing of the first sample until mid-November and the programme continued in a modified form until the end of February 1978.

Manufacturing problems arose during production of samples and, because of these difficulties, none of the samples was exactly as designed. As a result, the test programme was not so conclusive about performance as had been hoped. However, it gave enough evidence to justify predictions of a substantial improvement. In addition, a cost estimate of 'BU' product showed a reduction of £164 from the existing product cost of £716.

Based on this evidence, Mr Jenson, head of 'A' division, became involved in decisions on further action. He made a characteristic, autocratic, and emphatic judgement: he decided that the only way to prove the design adequately was to test it as a full-scale unit in the field. In March 1978 he issued a memorandum requiring that arrangements for producing units ready for field trials in September 1978 be made, and also that Mr Watkinson concentrate all his efforts on achieving results that would enable decisions to be made on the future of the 'BU' design during the 1978/9 financial year.

A programme was drawn up to meet this requirement involving four main aims:

(1) To transfer knowledge already gained by the product development department (PDD) to the process engineering department (PED);
(2) To agree a method of manufacture of the 'BU' block between PDD and PED;
(3) To get the applications department and sales to contact a suitable customer for agreement to field trials;
(4) To manufacture, assemble, and test a unit.

The first three aims were relatively straightforward, although the second introduced some changes from the original sample manufacture and it was felt essential to prove these by two further samples before assembling a full-size product. This, it was agreed, should be put on trial at Obah in the Middle East in place of the existing unit.

The manufacturing changes, which were intended to overcome the difficulties encountered during the earlier sample test programme, involved different operations. In the longer term, these would involve design and manufacture or purchase of additional equipment, but Mr Jenson firmly rejected any such expenditure in the immediate future with a memo stating 'Up to satisfactory conclusion of the first phase of field trials we must limit process investment to virtually nil'. His original request for field trials by September proved not possible when outside deliveries were found to be a limitation. A critical path network then indicated that the prototype product could be installed during November 1978.

The changes in manufacturing techniques and assembly times which the development experience indicated at this stage caused a fresh estimate of cost to be made. The effect was to raise the overall cost for a 'BU' module to within £17 of the existing unit.

At the same time the results of further tests of standard units became available and indicated that the performance predictions may have been underestimated. In actual practice, it appeared that the measured performance of the 'BU' design varied between 5 per cent and 9 per cent lower than standard over a range of conditions, where the planned improvement was 35 per cent.

By the beginning of October 1978 all the foregoing facts had become evident and it was clear that the original advantages claimed for the 'BU' design were not going to materialize without a great deal more development effort and expenditure. Mr Bales assessed the position and decided that there was no justification for proceeding. He therefore suspended all further work except for the preparation and issue of a full report on the work already carried out.

8.11 'Sink or Swim' Project

This project was aimed to re-establish the company's presence in the automobile industry but, because of the firm's lack of recent experience in the small-vehicle field, it had no first-hand knowledge of the improved design techniques and it realized that even more of its market would be lost unless the growing penetration of competitors with improved designs could be overcome. It decided, therefore, that a major effort was required but that initially this should be confined to a single product. It would not attempt radical design advances but be confined to marginal improvements over competitors' products. The argument for this approach was that customers were very conservative and were more likely to accept an evolutionary product, and that any problems could be solved and lessons learned on a restricted production. In addition, it would reduce the target time to production and enable more accurate forecasts of cost to be made.

The first formal step towards this was taken by Mr Jenson issuing a memorandum in April 1978. It initiated a programme which involved the whole of 'A' division in an immediate effort to achieve a new component. His directive was 'to produce from pilot production facilities, a prototype of the most worthwhile application by the end of December 1978'. He indicated that this should have the highest priority and any conflict with other important work should be immediately referred to him. His memo continued with instructions on tooling and assembly requirements, on confining the project to a single-customer application, and of working to a budget which he established at £25 000. In addition, he laid down intermediate dates for defining the design, releasing working drawings, and ordering tools, and he allocated responsibilities to his departmental managers.

He was not interested in the project as an innovation but purely in matching or surpassing competitive designs and so gaining an economic advantage which would encourage the company's customers to continue with them. He believed that a great deal of manufacturing know-how could be gained by discussion with the tool manufacturers and that much of the necessary performance could come from a study of the competitive product. Finally, he wanted to prove the component before attempting to launch it over the whole market.

The new component was quickly named Besttype, to constrast with its competitor, called Newtype, and the first step towards its development was taken by Mr Dutton (head of process engineering) and Dr Powell (head of product development). They visited three specialist tool-making firms, two in

Germany and one in England, in early May with the object of gaining information on manufacturing equipment and processes and their relationship to essential product features of their design. All three companies were in broad agreement on the general tolerances and manufacturing requirements and their recommendations were mainly adopted.

As a result of the visits the firm of Jost of Düsseldorf was selected to provide a production machine, partly because it inspired greater confidence in its appreciation of the problems and partly because it promised to supply prototype parts ahead of the company's tool-availability by using another machine tool before its delivery elsewhere. A British firm was also invited to quote for modifying one of the company's machines originally supplied by it in order to produce an essential component part.

While decisions on these production requirements were being made, work proceeded on designing the dies. Following the policy-directive against major changes, these were generally similar to those of Newtype but aimed to give modest improvements in performance which would, in total, add up to a significant overall improvement. At this point it was learnt from the machine-tool suppliers that competitors were moving in a new direction that would lower the weight and cost of their products. A design modification was immediately introduced which was later to create strength problems.

At this point, an analysis of confidence level in Besttype was made and, after allowing for possible inaccuracies (such as the performance predictions being in error and the Newtype measured performance not being typical) it was concluded that in the most pessimistic conditions Besttype would have an equivalent performance and 11 per cent advantage in material cost. The way was, therefore, clear to commit funds and effort to the next stages, although Dr Powell seemed anxious to record all the detail of decisions and work on which they were based so that any future difficulties within the design or process engineering departments could not be passed back to the initial work of product development. In fact, he produced a comprehensive report which gave full details of the manufacturing constraints, performance predictions, and design by July 1978.

Mr Jenson continued to direct the project personally and was able to hold his staff to the target dates in the early stages. Among these it was intended to approach the selected customer in September with the project proposal to obtain their co-operation. The timing of this would be critical since, if the approach was too early, it might leak to competitors and if too late, it might be forestalled by them. Unfortunately, Mr Jenson left the company about this time and some slippage began to develop. The customer was not made aware of the proposal until November and the original target of a prototype by the end of the year also went back. This was partly due to delay in finalizing the German tool order and partly to late delivery of raw materials.

New target dates were scheduled, but much of the early pressure to achieve a quick result seemed to disappear. The parts ordered from West Germany were no longer expected to be received before the early weeks of 1979 and their assembly into the test-pieces was programmed for February. The orders, when delivered, proved to be below the specification requirements. They had become distorted and were undersize. There were other defects which would contribute to a poor performance and the parts could not be accepted. A visit to the suppliers was made in March to find the causes and to take action to correct the faults. The firm was able to explain the causes and readily accepted responsibility. After some attempts at rectification it agreed to provide new supplies at its own expense, but this incurred a further six-week delay. In the meantime, the rejected material was assembled for mechanical tests and performance. The results were, as expected, below requirements (by 13 per cent) but this gave encouragement to believe that the predicted performance would be achieved when parts to specifications were used. The new batch of material did not arrive from West Germany until mid–1979 and, again, proved disappointing. This led to concern about the production technique, which was still not to specification. It was even suggested that Mr Bales, who had taken Mr Jenson's place, should look for an alternative source of material while visiting the USA. At the same time, Dr Powell was asked to investigate how closely the existing prototype could be matched to the Newtype performance by modifications in certain physical properties and whether economic advantages could still be claimed. From exhaustive test measurements it was possible to deduce that the Besttype performance should be within 3 per cent of specification.

In early September it was concluded from this work that Besttype, as it currently stood, was marginally acceptable for general commercial vehicle applications. To test this conclusion Mr Bales ordered a programme to manufacture and test samples which would, if successful, be despatched to the selected customer. He reintroduced some urgency into his order by laying down target dates only 2 to 3 weeks ahead, but these seem to have been badly missed, as the first sample was delivered only on 31 October 1979. This proved to be satisfactory from the customer's point of view, as they were just about to start a test programme with other units, and the insertion of Besttype into this programme would enable an immediate and direct comparison to be made. The test was carried out three weeks later and proved that the new product was at least as good as and possibly marginally better than its rival.

Following this performance Besttype was accepted and it was expected that negotiations for future supplies would be arranged with the sales department. There was still, however, only a slow preparation for production. Production engineering was making tools, the West German machine was to be delivered in March 1980, and a subsidiary machine was arriving from a British firm in the following October.

8.12 Conclusions

After approximately two years this study of the company's operations was completed early in 1980. During this period the progress of three company projects was followed. All three were undertaken against a background of past restrictions on innovative work and of fighting a losing battle against entrepreneurial competition. A change of attitude in 1977 had occurred and hopes were raised for a recovery of morale and technical enterprise.

The projects were clearly ones which required different periods of time to reach useful results, but the approach to them was largely the same. Only in part of the 'BU' project was a formal planning network drawn up, despite the fact that Ashworth House commenced each project by using procedures to locate uncertainties and assess risk. Indeed, work was scheduled in a series of short-term targets (a few months). This did not provide very coherent progress and tended to stretch out the time-scale and raise the expenditure. The latter was usually underestimated and not well controlled. The consequence was that all the projects took longer and cost more than expected.

The staff working on the projects were capable technical people, but their loyalties were often divided and their spirits dampened by the magnitude of their tasks. This was not helped by the unsettling effects of management changes brought about by key personnel leaving.

The net result was that the three longer-term projects overran their target dates and, after two years, the only definite decision was a negative one to suspend further work on the 'BU' project. Neither of the other two had reached a state where the effort expended had produced a useful return and they were being continued in a somewhat desultory manner. This discouraging state was not aided by subsequent events, i.e. the financial loss of the company during 1978–9, a pull-out of Ashworth House from the firm's commitments, difficult cash-flow problems resulting from the national economic climate, and further losses of staff. Happily the shorter-term prospects were more successful and all the six groups were completed within approximately eighteen months.

Ultimately, in July 1980, although beyond the period of this study, it may be recorded that the company was sold by the group.

174

Appendix I Company Organization Chart (1978)

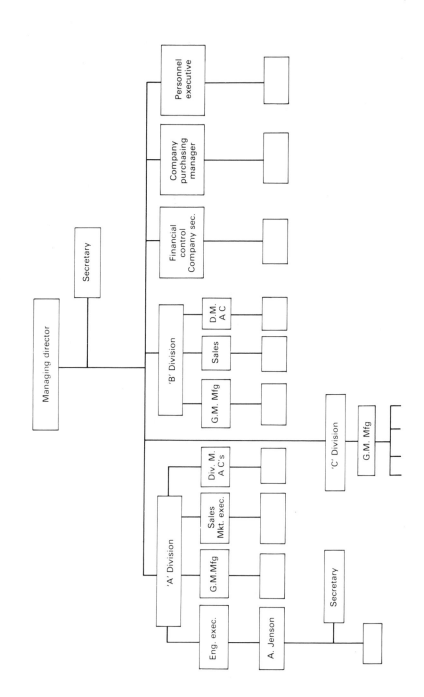

175

Appendix II   Executive Engineer—'A' Division: Organization Chart (1978)

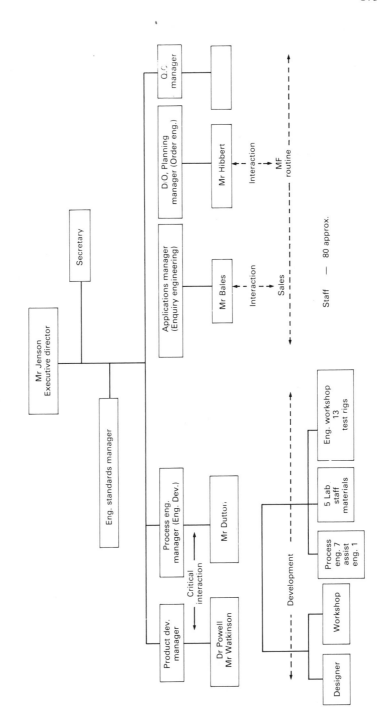

CHAPTER 9

An Engine Manufacturer (Case History No. 7)

When a firm that has enjoyed an outstanding export performance for over half a century is taken over by a large multinational group of international renown it would be reasonable to expect a favourable synergistic interaction. This case history describes a merger when this did not happen. The company had a full short-term order book, was complacent, and was inclined to avoid expenditure on future activities that might have prejudiced the attainment of the financial targets that were set by the parent board. Meanwhile, competitors, motivated by a severe downturn in world trade, were developing new products.

9.1   Introduction

This company has been established for over a century and has been producing combustion engines for 75 years, and a specialized range of engines for over 50. This case traces its developments both inside and outside a multinational corporation, considers its strengths and weaknesses, and looks at the current situation and its future options. The interest of the case thus centres on a subsidiary company and its parent group, both of international renown, and who nevertheless need to react to a worldwide economic recession. In particular, how do those who direct the company's affairs decide on the deployment of resources to anticipate future challenges?

9.2   The Organization

The organization is headed by the managing director who reports to the managing director of a multinational group making engines. There are five departments headed by the technical director, the commercial director, the manufacturing director, the personnel manager, and the chief accountant/company secretary. At the next level the training manager reports to the personnel manager, the production manager to the manufacturing director, and the chief engineer design and development and the chief applications engineer concerned with installation, fuel systems, ancillaries, and spares to the technical director. Some details of senior staff are given in Appendix I. The company employs 2000 people at two sites. The distribution of employees into the various activities of the company are shown in Appendix II. The overall company employs 200 000 at 200 sites, so that this firm is a 1

176

per cent factor in the overall structure. The company's balance sheet for 1978/9 is shown in Appendix III.

Approximately 50 per cent of the business is in the export field. This varies between 40 per cent and 60 per cent, dependent on the balance of orders at a particular time between different applications and customers.

9.3 The Current Situation

There is a full order-book for the next two years. The engines are highly regarded internationally, as use by nearly fifty countries illustrates. So in the short term commercial prospects appear to be satisfactory. Nevertheless the situation merits closer inspection. Its international competitors have a wider range of products. They also have a greater opportunity to operate in consortia to provide packages. Against this there is a trend towards smaller and faster vehicles, and a potentially quite large market for small light-weight engines. Assessment of whether making a concentrated effort towards moving into this market is viable depends on an intimate knowledge of the degree of segmentation and the potential size of each segment. This requires market research. In-house there is a good sales intelligence but no market research, and market research at group level for a range of firms may suffer from remoteness.

Research and development is also centralized. The relationship between a centralized laboratory and a firm is that while the firm is autonomous in so far as responsibility for its R & D is concerned, in practice this firm spends less than 2 per cent of its turnover in development and has no in-house research. The central laboratory determines its programme in terms of best use of its resources, and individual firms are free to use these resources or to choose other organizations outside the multinational company. This firm makes little use of central laboratory resources but uses polytechnics and consultants. Moreover the annual report of the holding company, i.e. the multinational, for the year ending 31 March 1979 makes no mention of research in the area of manufacture in which the firm is engaged. Whatever the merits of the commercial relationship deriving from each firm in independent profit-creating centres, one effect of this arrangement is low employment by the firm of young graduates who could be expected to be the source of innovative ideas.

As a result, in the post-war period developments have been almost exclusively of the product-improvement type. The strongest trend has been towards higher power output (BHP) and improved power-to-weight ratio. This is consistent with an effort in the smaller-vehicle market but this would involve capital expenditure in manufacturing plant. This is because current machinery is unsuitable for the purpose and extensive modernization would be required.

There are other plant-limitations. The workshop is laid out on a functional basis rather than in project cells. This involves across-the-floor movement of items and liaison between different groups of employees, placing limits on productivity. The ratio of engine testing time to production time is exceptionally high. It is recognized that operation to maximum production capacity is the key to profitability but there is less certainty about whether maximum production capacity should be defined against current planned and organizational arrangements or what it could be, given improvement in resources. There is currently a good deal of subcontracting. It is arguable whether this is to accommodate deficiencies in plant and organization or for the normal reasons that there are some items which can be manufactured more economically and flexibly by small specialist firms or to meet infrequent peaks. This uncertainty leads to *ad hoc* arrangements rather than a sub-contractual policy; this in turn is only justifiable in the short term pending improvements in resources.

Industrial relations are not a serious obstacle to changes in organization or plant. There are nine unions. Effectively there is a closed shop for blue-collar workers. Strikes are not unknown but they are neither frequent nor persistent.

9.4   The Group Structure

The multinational company in its present form was created through the merger of three companies a decade ago. It is a large organization with 200 000 employees organized into groups of companies manufacturing products in a particular field. The aim is for each individual company to be a profit-making organization in its own right. To this end it is autonomous in so far as it is free to use or not use centralized services. In practice, this firm buys in only 20 per cent of its raw materials and components from other companies within the multinational. Firms have a good deal of freedom in recurrent expenditure and employment of staff but do not retain profits and require approval for modest capital expenditure.

There can be little doubt that the merging of the three manufacturers offered an opportunity to increase profits by the elimination of duplication and the availability of higher capital investment. Overall this has been proved in the first decade of working. In round terms while the retail price index has doubled, sales have trebled and profits increased by 6:1.

For firms with steady growth and expanding markets these arrangements appear to approach the ideal, but for areas where existing markets are retracting the balance of advantage between autonomy and independence to invest in modernization of equipment or engage in directly relevant market research is much less evident. Put another way, the inherent policy appears to be much more beneficial when there is no urgent need for new markets and new product ranges or significant increases in productivity to retain a firm position in existing markets.

9.5   The Way Ahead—The Options

Option 1

The first and easiest option is to continue along present lines against a current full order-book, an overall policy of short-term profit, and some policy restriction against substantial capital expenditure of benefit in 5–10 years' time. This has some attractions. It is, however, very dependent on the reputation of the firm's engines and lack of additional muscle by their competitors if the market diminishes.

Option 2

Modernize plant and improve organization of workshops to improve both productivity and the ability to compete in new markets and/or to hold position in existing ones. There have been encouraging signs where improvements have been made, e.g. with new machinery the lead time for the product has been reduced from 36 to 6 weeks. An important and promising study was undertaken in 1978. It forecast that the organizational and technological changes arising from modernization of plant and a change-over to project cells would result in lead times being reduced to 15 per cent at current levels, reduce the direct labour per output to half its present level, and significantly reduce unit costs. There might be a further opportunity to integrate implementation of this with a major building and development programme.

Clearly, the benefits arising from such changes are dependent on an accuarate and directly relevant market research effort and the assistance of staff capable of innovative effort to capitalize on new advantages.

Option 3

The acquisition of a new company or the use of existing subsidiaries to diversify to new products or applications. The absorption of a part of another firm ten years ago is an encouraging precedent, as ten years later this accounted for a significant percentage of the firm's total business.

These options are not mutually exclusive. For well-based decision-making on the choice between them, senior management needs costings for modernization to increase product range and productivity and comprehensive marketing data. Given this, a corporate plan with growth targets might be developed to assist in determining priorities to minimize risk in decision-making and to point to innovations which will give rewards.

Appendix I Board of Management

The board of management consists of the managing director and five heads of departments — the technical director, the commercial director, the manufacturing manager, the personnel manager, and the chief accountant who is also the company secretary.

The managing director has been in post for seven years, and is in his mid-fifties. He came to the company from another engineering company where he was financial director. He is an accountant by profession and all his training and experience has been in accounting within the engineering industry.

The technical director has held his appointment for thirteen years. He is in his late fifties and his background has been in the design and development and project fields in engine companies. He is a PhD and a Chartered Engineer.

The commercial director has been in post for six years, after being a marketing director with the multinational company. He was an engineering apprentice with HNC qualifications. He is in his late fifties and has spent twenty years in the sales side of the business including a period as a resident representative overseas.

The manufacturing manager is in his mid-fifties and has been with the company for thirty-two of the last thirty-five years and has been in his present post for twelve years. He was an engineering apprentice, and most of his experience has been in service departments specializing in quality control. He is a Chartered Engineer.

The personnel manager has been in the post for five years and is in his early fifties. He has spent fifteen years in personnel departments in the pharmaceutical and rubber industries and immediately prior to his present appointment in a steel fabricating company. His qualification is MIPM.

The chief accountant and company secretary is in his mid-fifties and has held this position for ten years. All his training and accountancy experience has been within the engineering industry. His qualifications are FCCA, ACMA.

Appendix II Distribution of Employees by Activities

Manufacturing
	Direct*	606	
	Indirect†	791	1517
	Apprentices‡	120	
Administration and accounts			70
Technical			172
Selling			80
Service			77

Spares	74
Total employees	1990

*   Direct employees are those employed directly on production — 350 are skilled technicians.

†   These are ancilliary workers, e.g. crane drivers, lorry drivers, packaging staff, painters.

‡   This is a four-year apprenticeship intake (30 per annum). About 20% with four 'O' levels on entry are selected as technician apprentices for design, production planning, etc. They are taken academically to HNC or the TEC equivalent. Additionally three undergraduates are recruited a year. These normally have 'A' levels and are sponsored by the company for their sandwich degrees.

Appendix III   Balance Sheet 31 March 1979

	31.3.79	*31.3.78*
ASSETS		
Investments in subsidiary companies at cost		
(see note 4)	1 962	1 962
Loan to firm from holding company	3 568 245	3 566 064
	3 570 207	3 568 026
LESS: CURRENT LIABILITIES & PROVISIONS		
Subsidiary company balance —	10 435	10 435
	£3 559 772	£3 557 591
Financed by:—		
SHARE CAPITAL		
Authorized, issued and fully paid		
1 400 000 shares of £1 each	1 400 000	1 400 000
Reserves	116 242	116 242
Retained earnings	2 043 530	2 041 349
	£3 559 772	£3 557 591

SIGNED ON BEHALF OF THE BOARD

} DIRECTORS

REPORT OF THE AUDITORS TO THE MEMBERS OF THE COMPANY

In our opinion the accounts and notes set out in pages 1 & 2 which have been prepared under the historical cost convention give a true and fair view of the state of the Company's affairs at the 31st March, 1979, and comply with the Companies Acts 1948 and 1976.

....

Chartered Accountants
London, EC1N 8RN

Notes to the Balance Sheet 31 March 1979

1. The company has not traded during the year to 31 March 1979.

2. Group Accounts have not been prepared for the company which is a wholly owned subsidiary of another company incorporated in the United Kingdom.

3. Subsidiary companies — 100% share capital held in each of our two subsidiary companies.

4. Profit and Loss Account/Retained Earnings

A taxation adjustment resulting from an overprovision in previous years of £2181 was transferred through the company's loan A/C to profit and loss account and then to retained earnings.

CHAPTER 10

The Oxford Instruments Group Limited (Case History No. 8)

As manufacturing industries entered the 1980s it became evident that few of the large traditional industries could ever regain their former share of world markets. Economic growth would have to rely on the growth of smaller companies and particularly those whose staff exhibited a high degree of knowledge and skills: companies, moreover, whose sophisticated products are characterized by high cost per unit weight and a high added-value.

The Oxford Instruments Group, born 21 years ago, was chosen as being in the van of high-technology manufacture. The company, based on scientific excellence, has grown rapidly since 1973 and demonstrates that a first-class product will sell throughout the world even under the shadow of a depression. The company has gained three Queen's Awards: one for export and two for technical achievement.

10.1 Introduction

The Oxford Instruments Group Limited, like a number of other technological firms, was born in a university research department. The original company, Oxford Instruments, was formed in 1959 by the present group chairman, Martin Wood, when extramural services to academic laboratories led to larger orders from government establishments. Since Oxford Instruments became a limited liability company in 1961 it has grown to a group of four companies specializing in scientific, medical, and industrial instrumentation with a total annual turnover of £6m. The first order was completed by Martin Wood assisted by a retired technician. The average number of employees in 1979 was 500, one in three being a graduate and one in ten a PhD.

This case traces the developments over the last twenty years, leading to an organization engaged in high-technology manufacture and earning two-thirds of its revenue in the export market.

The interest of the case centres on the balance needed between freedom of action for highly qualified innovative people and the commercial exploitation of their results to obtain maximum financial advantage. The contrast between the 1960s and 1970s is striking in this regard. While a high standard of technological innovation was maintained throughout the entire period, little priority was given in the first decade to business matters such as personnel selection, salary levels, purchasing techniques, stock records, factory inspection, or the complications of importing and exporting. As a result, there were at least two financial crises when survival depended on the intervention of

183

external financial interests. In the second decade, when greater management skill achieved a balance between the commercial and technological aspects of the business, a period of growth occurred culminating in the significant financial advance over the 1978–9 period.

This case illustrates vividly that the frontier between bankruptcy and high profitability is as narrow as that between madness and genius. It also makes evident that well-manufactured, elegant, and uniquely useful products in advance of competition will sell anywhere in the world whatever the economic climate.

10.2 The First Steps (Pre-1965)

In the 1950s the Oxford University Clarendon Laboratory was developing magnets energized up to 2000 kW. The fact that these produced the highest magnetic field in the world caused many academic laboratories to seek help with their work from that laboratory. Martin Wood, a research scientist at the Clarendon, had already shown entrepreneurial leanings by constructing simple but ingenious medical instruments, an interest that was fostered by having a father and two brothers who practised medicine.

The first significant orders were for two large technically advanced magnets—one for the Royal Radar Establishment (RRE at Malvern) and the other for the Atomic Energy Authority (AEA). These were made in a garden shed.

Because sophisticated electronic power and water-cooling devices were required for laboratory work in a high magnetic field area, the market was restricted to some ten laboratories. For extension of the market, Martin Wood turned his attention to the medical field, where he had good contacts. To meet this, new accommodation was needed and an unused stable and slaughterhouse at Middleway near Oxford was purchased. This enabled Oxford Instruments to meet an AEA order for thirty-nine field-modulated coils. This and other orders led to the decision to become a limited liability company in 1961.

A conference at the Massachusetts Institute of Technology (MIT) on high magnetic fields in November 1961 proved to be an important milestone in the development of Oxford Instruments. At that time, the phenomenon of superconductivity, i.e. the disappearance of electrical resistance at temperatures approaching absolute zero, was a 50-year-old interesting phenomenon of no known practical use. Discussion with other researchers in this field at the conference inspired Martin Wood to decide that Oxford Instruments would manufacture superconducting magnets. The potential advantages of this decision were immense. It would make available high-strength fields to laboratories lacking power plant and the water-cooling necessary for conventional magnets. This would replace a market of ten laboratories by one of vast potential. It was a development, too, which involved low capital outlay to produce compact low-weight magnets. The only significant problems were dependence on external sources for an adequate supply of liquid helium and wire of suitable material for the magnets. It was a bold and imaginative decision.

The first magnet used niobium-zirconium wire produced in the USA at a higher cost than gold. Ready for testing in April 1962, it produced an intense field, could be held in the hand, and could be driven by a car battery. Orders quickly built up. Additional machines were bought second-hand from government surplus. Growth of orders led to the need for full-time staff. Six, including Frank Thornton, a physicist who later became managing director, were recruited in 1963 and, by 1965 the staff grew to twenty-five, almost all of whom were engineers, scientists, and technicians.

Unfortunately, the initial trouble-free experience in manufacturing magnets dependent on the phenomenon of superconductivity did not continue. Later magnets proved to be variable, unstable, and prone to damage. Shortage of liquid helium also caused concern; supplies were neither adequate in quantity nor low enough in cost.

To relieve the latter problem a separate company, Oxford Cryogenics Limited, was formed. Financed by a bank overdraft, a semi-automatic Collins Helium Liquifier was purchased from the USA. This served to supply Oxford Instruments Limited and sell 3000 litres in the first three months. Oxford Cryogenics Limited served a useful purpose for three years or so, but it led to the British Oxygen Company, the company's original main supplier, increasing its production and starting a price war. As a result, Oxford Cryogenics was sold at a small profit in 1968. The BOC collaboration, however, continued and some five years later the physical resources of BOC's superconducting magnetic system division was transferred to Oxford Instruments in exchange for a 15 per cent equity holding.

10.3   the Middle Years (1965–72)

Despite the problems, Oxford Instruments Limited continued to expand and, in 1965, a 3000-ft^2 boat-house on a two-thirds of an acre site by the Thames at Osney Lock was purchased. It was re-equipped and occupied in six months. An adjacent one-third of an acre was also purchased to provide for further expansion.

The years 1966–7 brought new financial and technical problems. The US suppliers replaced niobium-zirconium wire with niobium-titanium. Some magnets unaccountably failed to achieve specification and others failed on delivery. Projects were held up for as much as six months and the cost of investigation increased dramatically. As a result, in 1963 accounts had to be withdrawn before publication since stock could not be reliably valued.

Fortunately Oxford Instrument's troubles were also experienced by their US competitors, some of whom went into liquidation. As a result, the company weathered the storm during the famine of usable superconductors until a UK company, Imperial Metal Industries Limited, produced new and improved niobium-titanium wire. The change-over to IMI wire during an expanding phase in the world's economy combined to give good trading figures. In April 1967 the company received the Queen's Award for Technical Innovation in recognition of its manufacture of the strongest magnets and of the coldest refrigerator in the world.

Nevertheless, finance was now a matter for serious concern. The company bank became restive and disaster was avoided first by the Midland Bank taking over the company overdraft and debentures and so providing breathing space for negotiation of longer-term finance. This came from the Technical Development Corporation (TDC) who, having regard to the trading figures, the Queen's Award, and the Ministry of Technology's enthusiasm about the 'white hot technological revolution', bought 20 per cent of the equity in special shares bearing a fixed dividend and put a similar sum in loan capital to a total of £90 000.

In an atmosphere of greater financial stability, the late 1960s were notable for more substantial technical advances. Dilution refrigerators were built operating below the boiling point of liquid helium within a fraction of a degree of absolute zero. Superconducting magnets with very uniform fields were developed for nuclear magnetic resonance spectroscopy. The Ministry of Technology financed development work on powerful superconducting quadruple beam-focusing magnets for the European Nuclear Research Centre (ENRC) at Geneva. A standard cryostat was designed for application in X-ray diffracrometers and infra-red spectrometers.

In summary, the first decade produced such a flow of significant scientific and technological achievements that tiding-over finance was available and masked the fact that management and commercial skill did not match technological accomplishment. This uneasy balance was not to continue. New technical problems produced a further financial crisis.

Superconducting wire was again at the root of the problem. Unsatisfactory performance led to delivery delays and cash-flow problems, and the bank wanted to appoint a receiver. The International and Commercial Financial Corporation (ICFC) then entered the scene and provided loans conditional on a review of the management of the company. As a result, Frank Thornton and some seven senior and middle managers left. A case for an experienced and more professional managing director was clear and ICFC's personnel selection department advertised the post and provided a shortlist of candidates. The outcome was that, in 1970, Barrie Marson was appointed managing director. He was well suited for the post. As a director of Kent Instruments he had built up their digital systems division from scratch, and, moreover, trained as a physicist, he could easily communicate with his scientific staff. The departure of seven managers enabled Marson to bring in replacements with wider management experience. So a new era began, heralded in 1972 by another Queen's Award—this time for export: by now exports were running at between 60 and 70 per cent of output. But although small losses continued in that year profits grew rapidly from 1973.

10.4 The Last Years (1973–80)

Over the last eight years there has been a balance between technological achievement and management skill. It has been a period of growth in more often than not a difficult economic environment. The technical lead in superconducting magnets has been maintained, although not without difficulties. A fall in orders for magnets, which began in the early 1970s, was partly due to spectrometer manufacturers developing their own magnets. One of them, recognizing the unique experience of Oxford Instruments, made a take-over bid. Although the offer had some attractions, it was decided to retain the company's independence and make strenuous efforts to gain a technical lead by a new advance. The opportunity came from the need by Oxford University for a high-field stable magnet with resonance frequencies up to 470 MHz. The difficult technical problems were solved. The new magnet had the advantage of needing to be filled only every three months. The order book grew to £1m in a short time. It won the coveted R100 Award from the USA. Orders of this magnitude involved subcontracting, and two additional buildings had to be acquired. The scientific and medical press has also recently drawn attention to new advances in medical imaging by Nuclear Magnetic Resonance (NMR), which has been made possible by Oxford Instruments developing large-bore magnets that allow access to a patient's head or limbs. In Appendix I is reproduced an extract from the *Sunday Times* (1980) on the medical application of this technique. In 1979, all activities were brought together at Osney.

In 1979 a new company, Oxford Research Systems, was formed with the object of developing commercial, medical applications of Topical Magnetic Resonance (TMR), which is a technique for obtaining high-resolution NMR and constitutes a new harmless non-invasive technique for studying *in vivo* the metabolic state of internal tissues. Unlike NMR, TMR does not produce a spatial image. Its purpose is to identify diseased tissues, and current work is directed to a study of phosphate metabolites on tissue. (See Appendix II for an extract from the *Physics Bulletin* (1980).

Diversification has also been actively pursued. The outcome of the development of a miniature tape recorder for monitoring heartbeats by Dr Barry McKinnon, the research manager of Oxford Instruments, was a Medilog Recorder. This led to the formation of Oxford Medical Systems Limited at Abingdon. In 1974, Newport Instruments of Newport Pagnell became, for an exchange of shares, a wholly owned subsidiary of the Oxford Instruments Group. At the time, Newport Instruments had three divisions concerned with instruments, electronic components, and caravans. The caravan interest was sold off and the remaining activities transferred to Milton Keynes.

In 1974 a relationship developed between the Oxford Instruments Group, Newport Instruments, and the Italian company Praxis. This latter company was soon to be taken over by the Italian instrument group, Carlo Gavazzi (see

Appendix III), when one result was a decision to manufacture the Gavazzi system at Newport Instruments' factory at Newport Pagnell. A small team of systems engineers and specialists was recruited and this became a new division of Newport Instruments under John Lee as managing director in 1977.

So the 1970s saw a continuance of the existing technological advance which occurred in the 1960s, but with the difference of greater financial stability, diversification, and rationalization. The group enters the 1980s with confidence, given a high export ratio despite higher-than-average UK inflation and the strong pound, and the financial improvement revealed in the 1979 Report.

10.5 Organization

The Oxford Instrument Group comprises four companies: Oxford Instruments Limited and Oxford Research Systems at Oxford; Oxford Medical Systems Limited at Abingdon; and Newport Instruments Limited at Milton Keynes. The group board has six directors: Martin Wood is the chairman, and Barrie Marson the managing director.

Oxford Instruments Limited specializes in cryogenic systems for scientific purposes and they are world leaders in the manufacture of extremely powerful superconducting magnets. The managing director is John Woodgate; other directors are John Pilcher, production director; P. Hanley, technical director, M.F. Wood, chairman; G.B. Marson, group managing director. (An organization chart appears in Appendix IV.)

Oxford Medical Systems Limited provides medical instrumentation and physiological health-care monitoring systems. It has a world-wide reputation, particularly through the Medilog system for cardiac monitoring. Its managing director is John Lawrence. Other directors are D.N. Reeve, A. Costley-White, and Dr J.R.W. Morris. (An organization chart is shown in Appendix V.)

Newport Instruments Limited provides the industrial side of the group's activities. It has occupied a new 20 000-ft^2 factory at Milton Keynes since December 1978. All its manufacturing was then brought together from five sites. Traditional products are a wide range of wound magnetic components and pulse transformers and a nuclear magnetic resonance process (NMRP) analyser for use in the food, petrochemical, and allied industries.

In 1976 the company entered the microprocessor business with distributed data and acquisition and control systems. Its managing director is John Lee. Other directors are J.R. Thomas, production director, and R.C. Milward, the technical director. (An organization chart is shown in Appendix VI.)

Oxford Research Systems Limited has been established to provide a directed effort towards new development. The group believes that it is necessary to invest about 15 per cent of sales receipts to ensure the future by continuing to introduce new and advanced products ahead of competition. To this end, regular contact is made with twelve universities and medical organizations around the world. This is because experience has shown that ideas for new products arise mostly from customers.

A new NMR spectrometer is under development. The company has invested £500 000 and additional finance has been provided by NRDC. A negative cash-flow is anticipated for the first 2–3 years.

The company is housed in leased premises within walking distance of Oxford Instruments Limited. Peter Hanley, formerly development director of Oxford Instruments, is the managing director.

Two-thirds of the group's turnover is in overseas sales and a strong overseas presence is maintained with subsidiary companies in the USA, Australia, West Germany, France, the Netherlands, and Switzerland. In addition, there is a world-wide network of distributors.

There are 500 employees, including 100–150 graduates and 40–50 PhD's. Raw graduates are not recruited. Postgraduates from university or industry are obtained when management skills are required. There are no unions.

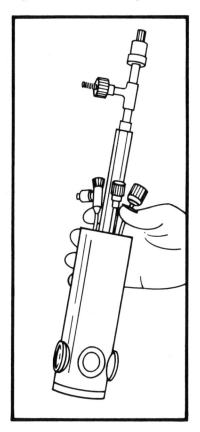

Figure 10.1   CF104 cooling unit

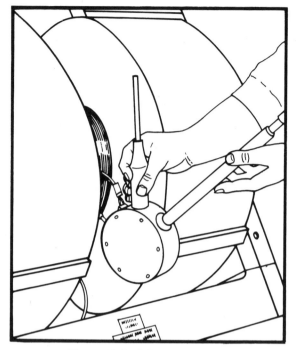

Figure 10.2   ESR900 system

In Appendix VII is given brief biographical details of the chairman and managing director of The Oxford Instruments Group Limited, and the managing directors of Oxford Instruments Limited, Oxford Medical Systems Limited, and Newport Instruments Limited.

10.6   Product Range

Oxford Instruments Limited's wide spread of products includes cooling units, superconducting magnets, and ultra-low-temperature equipment. Examples of two cooling units are given in Figures 10.1 and 10.2. The range of cryostat systems for low-temperature research comprises a series of bath-type cryostats: the MD series for liquid helium and the DN series for liquid nitrogen.

The company has been manufacturing superconducting magnets for nuclear magnetic resonance processor analysers (NMR) since 1968. Figure 10.3 contrasts a simple and more complex magnet arrangement. Oxford Instruments was the first company to produce commercially 270-, 260-, 470-, and 500-MHz magnets. Additionally, cryomagnetic systems providing magnetic fields up to 8 tesla are produced for spectroscopy and a full range of magnetic susceptibility systems for many applications in physics and chemistry. (See, for example, Figure 10.4) A comprehensive range of

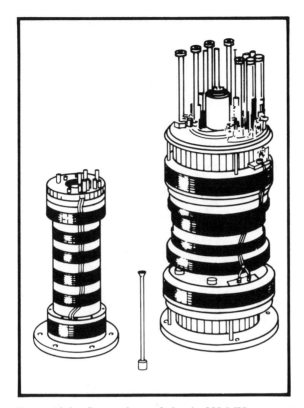

Figure 10.3   Comparison of simple 220-MHz magnet
and complex 500-MHz magnet

Helium-3 systems and dilutionary refrigerators is manufactured for experiments in the temperature range below 1.5 K. (A list of products is shown in Appendix VIII.)

There is a back-up world-wide service, 'Cryospares', for the rapid supply of various materials and ancillary equipment used in low-temperature research. This service includes supplies which are normally difficult to obtain quickly or in small quantities.

The Medilog Ambulatory Monitoring Recorder and Analyser systems contributed 90 per cent of the output of Oxford Medical Systems Limited during 1978–9. The Medilog products were launched more than five years ago and are facing increasing competition, mainly from more recent US alternatives. To counter this, a completely new Medilog Recorder technically in advance of current competition was introduced in the USA and West Germany in 1979. Additionally, a new multi processor-based scanner, intended for the markets of both the existing Medilog Analyser and a larger low-cost market in the USA, was introduced at the same time. Besides the two varieties of Medilog the company manufactures EEG analysers, page-

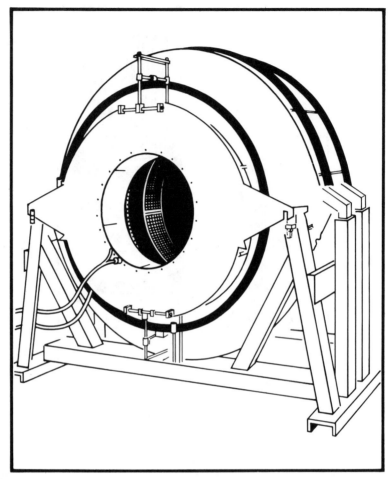

Figure 10.4   A whole-body NMR magnet: weight 3.6 tonnes; bore 616mm

monitor displays, and bedside computerized monitor/alarm systems. Figure 10.5 indicates a Medilog-2 Analyser. (The product range is shown in Appendix IX.)

Newport Instruments Limited has a number of traditional products including a wide range of wound magnetic components and pulse transformers used in OEM electronic applications. Wound magnetic components have been manufactured for twenty years and supplied to a broad cross-section of the world's electronic engineering industry. There are some ninety customers drawn from the public corporations and large and small industrial firms. Where standard ranges do not specifically meet customer-requirements, free design service is offered. Typical current applications are to be found in the following fields: thyrostat triac drives; data processing;

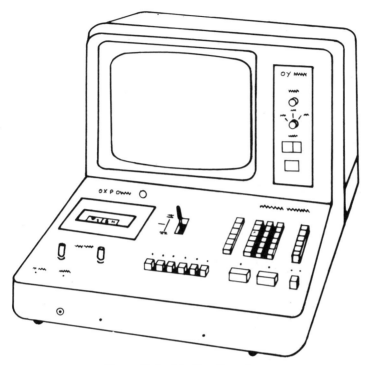

Figure 10.5   Medilog-2 analyser

communications; airborne computers; scientific equipment; general instrumentation; industrial electronic controls; motor control systems; computers;
nuclear power stations; medical electronics; guided weapons; and isolation
circuits.

The company also developed a fast protein-analyser for use in the food and
allied industries. Based on a pulsed NMP technique, it enables the direct
percentage by weight of protein in a prepared food sample to be determined
in 20 s.

In 1976 the company entered the microprocessor business with a powerful
distributed data-acquisition and control system for industrial and process
applications based on the Praxis DIAC system. The first system was installed
at Tilbury to monitor and control the world's first refrigerated container-
terminal to be remotely controlled by computer.

In 1977 the company's position was strengthened by substituting their own
System 86 for the DIAC system. A bold decision to draw up product-
specifications incorporating CMOS (Complementary Metal Oxide Semi-
conductor—a distinguishing feature is its low power consumption) ahead of
its commercial availability resulted in their marketing a remote fifty-channel
multi-plesser unit with a power consumption of 1 W and therefore capable of
being used in an hydrogen-explosive environment. (Appendix X shows the
main products.)

It has been the experience of the manufacturing companies in the group that ideas for new products mainly originate outside the company. This parallels the convictions aired at Oxford Research Systems and possibly explains why technical meetings with representatives from each company were not found to be worthwhile. Senior staff variously commented that

> ...all ideas came from world centres of excellence. We have to be good at interpreting customers' ideas since, though many are silly, a few are excellent. We have no problem in finding ideas, since customers come knocking at our door with them.

Unusually, their development programmes were completed in two years and successfully launched within a further six months.

10.7 Finance

The growth of the group over its twenty years of life is shown in Appendix XI. This provides a continuing annual picture of financial turnover and staff members. It indicates quite clearly the improvement in the turnover/staff numbers ratio since 1973 and that the poorest ratio in the middle years resulted in a financial loss over the 1969/72 period.

The turnover of the group in 1979 was £5.8m, as compared with £3.7m in 1978. The export proportion was £3.6m and £2.1m, respectively. Profits grew from £402 249 (£205 429 after taxation) in 1978 to £804 536 (£532 003 after taxation) in 1979. Total assets rose from £1.2m to £2.1m in the year ended 25 March 1979. In common with other export-dependent UK companies, the strength of the pound has put pressure on both competitiveness and profit margins in overseas markets.

The Consolidated Profit and Loss Account for the year ended 25 March 1979 is shown in Appendix XII and the Consolidated Balance Sheet at 25 March 1979 is given in Appendix XIII.

10.8 The Current Situation and The Way Ahead

As the financial statements indicate, the group entered the 1980s in a period of growth. As compared with 1978, 1979 turnover increased by 57 per cent, exports by 71 per cent, profits before tax by 100 per cent (after 159 per cent), and assets by 75 per cent. Continuation of this pattern is largely dependent on external economic influences.

Successful UK exporting firms are faced with a strong pound and UK inflation rates at a much higher level than in many countries to which it is exporting, e.g. four times that of West Germany. To offset this, retain competitiveness, and accrue satisfactory profit margins, it is necessary to obtain sharp reductions in unit cost through improved efficiency or productivity, or much higher sales. Present policy is to leave dollar and deutschmark prices almost unchanged, but this can only be a holding operation against the hope that present government monetary policy will produce relatively early results.

Without that, continuing success is highly dependent on innovations and advanced new-product development to provide products technically ahead of competition to generate new markets or to expand existing ones. This, together with improved production facilities and rationalization of product range to concentrate expertise, is perhaps the best option to tide over the difficult period until the economic climate improves.

Reference to the historical background section reveals that the record of the group in maintaining a product lead and in rationalization is impressive. The widening of the high magnetic field market by the introduction of superconducting magnets in 1962, the development of the dilution refrigerator in the late 1960s, and the Medilog-2 and System 86 in the late 1970s are outstanding examples of the continuing trend in new technological innovations in advance of competition. The phasing-out of nearly all non-medical products in the range of Oxford Medical Systems to concentrate on Medilog recorders and analysers, the selling off of the Coventry steel caravans by Newport Instruments, and the establishment of Oxford Cryogenics Limited in the mid 1960s to counter the production limitations by shortage of liquid helium are interesting examples of rationalization and improvization to ensure that strategies were fulfilled.

The successful continuance of the strategy to provide a flow of commercially sound technological innovations on a 'first in the field' basis demands high calibre and imaginative management. The high national unemployment is currently helpful but, given a return to something approaching full employment, competition for the best brains could prove to be a limiting factor on future growth. The company is now studying methods of employee-shareholding. If a suitable arrangement can be found within the possibilities of new government legislation on this subject, less wastage of good people would be probable. Additionally, such an arrangement would have the financial advantage of some shifting of reward to employees from a charge on costs to a charge on profits.

The track record over the last twenty years in obtaining a lead and sustaining it must be cause for the group looking to the future with optimism. Not only has there been an ability to make a technical advance when competition has become ominous but there is also a capability of overcoming teething problems in a new field with little or no experience to draw upon, at least as quickly as competitors. The combination of inflow of new ideas and commercial application owes much to the close links with Oxford and other universities and contacts with the Medical Research Council.

The basic elements of the group's strategy are scientific excellence, an emphasis on technically advanced products, and a concentration of effort in fields where special expertise exists to maximize the likelihood of keeping ahead of competition. To achieve this, it is a primary objective to recruit the best brains and keep them.

Whereas most companies require a forward corporate plan not only to stimulate innovation but to decide between options, a corporate plan does not appear to be relevant to the Oxford Instruments Group for as long as the present strategy continues to produce results. Indeed, the only attempt made so far at a corporate plan was misleading in that the results exceeded forecasts. Long-term high-investment planning, except perhaps in areas such as the systems manufactured by Newport Instruments, is not as valid a reason for forward planning as, for instance, it would be to an aviation firm.

What, then, are the factors which might endanger the continuance of present strategy? What steps can be taken to limit these threats? The two main dangers appear to be external economic factors and a reduced ability to recruit the best brains if the supply/demand ratio for highly qualified technologists and managers became less favourable. The strong pound and above-average inflation minimize profit margins in the export field and high borrowing rates limit investment capability in new and advanced developments to maintain leads over competitors. As has been said, this problem can be alleviated either by reductions in unit cost of production or by increased sales, or both. The group chairman in the 1979 Annual Report indicated the company's awareness of this when he said:

> ...our main objectives during the year were the provision of improved production facilities and a continued high level product development.

Rationalization by exclusion of peripheral production outside the field of greatest expertise has already been largely effected. The elimination of non-medical products in the Oxford Medical System range and the selling-off of the Newport Instruments caravan interest are examples of this. The current investigation of employee-shareholding possibilities, if brought to a successful conclusion, may also serve to retain the best brains.

Two areas of activity which may need to be developed further are the gathering of intelligence as an aid to new development and market research to keep under continuous review forward priorities on the balance of production effort between the different ranges of output.

Newport Instruments Limited might serve as a pilot case on this latter consideration. It is understood that their present analysis is that components will grow at the rate of inflation, but there is strong competition at the low-price end from Japan and Korea, instruments are a specialized and limited field and need world-wide markets for viability, and systems do not require research but derive from a combination of keeping abreast of the latest developments and reacting to customer-needs. Emphasis in the future should surely be on systems and on more specialized and sophisticated components such as those required for defence and the Post Office System X.

This no doubt takes the analysis a long way along the road, but systems require the highest investment at a time when external economic conditions at the behest of the government create low margins for a firm dependent on exports. There may be a need for a careful assessment of the timing of concentration of systems, especially having regard to the development/ lifetime being almost a 1:1 ratio.

Overall current success rests on the relatively simple strategy of good contacts, concentration on excellence of products and producers, and no dissipation of resources on peripheral products or broader planning than is necessary. This is not, therefore, a case for alternative distinct future options but rather of ensuring continuance of the good technology track record and achieving maximum benefit from it through a continuous review of product-priorities in market terms.

'Our financial aims,' said the group managing director, 'is a return on investment 10 per cent above inflation level and a doubling of profits every two years.' Given an early improvement in the international and national economic situation, Appendices XI and XII indicate a good chance of achieving this objective.

Appendix I 'X-rays Face New UK Challenge'

Extract from the *Sunday Times*, 7 September 1980
by Bryan Silcock, Science Correspondent

A team of scientists from Nottingham University has achieved a world lead in developing a new method for seeing into the human body. It has many advantages over X-rays.

At the British Association meeting in Salford last week, Dr Brian Worthington, the medical member of the team, unveiled some of the first high quality pictures obtained by the new technique. One showed a section through the head of a 22-year-old woman with a suspected brain tumour standing out clearly. With its help, the tumour was successfully removed.

Another picture was the first showing healthy vertebral discs ever to be taken from outside the body. Displacement of these discs is one of the commonest causes of back pain.

The equipment has been developed over the past six years by three members of Nottingham's physics department, Bill Moore, Rob Hawkes and Neil Holland. The quality of the pictures they have already obtained seems to prove they are well ahead of other groups on both sides of the Atlantic working towards the same goal: the medical exploitation of a phenomenon called nuclear magnetic resonance, or NMR.

NMR induces the nuclei of atoms in a strong magnetic field to emit brief, faint radio signals. In practice, NMR equipment is usually targeted on hydrogen atoms, found in nearly all body tissues. With suitable detectors and computer processing these signals can be assembled into a picture of a section through the human body.

NMR's chief rival is X-ray computerised tomography. This is a method of combining by computer X-ray scans taken from many different directions into a composite picture in which soft tissues as well as bones can be seen. It can even be used to look inside the skull.

NMR has similar applications but also has important advantages. It does not use potentially dangerous forms of radiation. It does not require contrast media — the chemicals that have to be injected to make some tissues show up in X-rays. It can take pictures directly of sections at any angle through the body. And by adjusting a few settings, the NMR equipment can produce pictures in which different tissues show up differently.

Negotiations are well advanced with a British company for the commercial exploitation of the equipment. It is bound to be expensive, but may cost less than that required for computerised tomography. 'With serious production, it might work out at about £100 000 a machine' said Moore.

Appendix II 'Topical Analysis'

Reference: *Physics Bulletin*, 1980. **31**, No. 9, 307. (Reproduced by permission of the Institute of Physics)

A new, harmless, non-invasive technique for studying *in vivo* the metabolic state of internal tissues has been put into practice by Oxford Research Systems, the development group of Oxford Instruments which is led by Peter Hanley and Derek Shaw. Based on conventional ^{31}P NMR, 'Topical magnetic resonance' (TMR) can be used to determine the biochemistry of a small volume of tissue, and thus should be particularly helpful in the identification of diseased tissue, in drug efficacy studies and in other medical research applications, despite being less sensitive than other spectroscopic techniques. In medical diagnosis it has the advantage that it can see biochemical changes as they arise, not just the symptoms of the chemical balance having gone awry.

The aims of TMR and NMR imaging are poles apart as TMR makes no attempt to produce a spatial image. TMR is a technique for obtaining high resolution NMR spectra from a selected volume of a larger specimen. The 'sensitive' volume is defined by superimposing high order magnetic field gradients, set up by higher strength versions of the conventional shim coils of wide bore superconducting magnets, to give a central 'uniform' field region surrounded by rapidly changing fields. The chemical shift range of ^{31}P is about 30 ppm and thus the central field region must be uniform to 1 part in 10 if the separate peaks are to be resolved. Such a requirement is feasible for fields up to 2 T. The rapidly changing field region gives rise to inhomogeneously broadened spectra within the receiver bandwidth, but these can easily be deconvoluted from the high resolution spectrum. Data collection times depend on which chemical (or range of chemicals) is being investigated and the level of detail required, but times of a few minutes for each spectrum are typical.

A role for which TMR seems ideal is the study of the major phosphate metabolites in tissue. These include ATP, phosphocreatine (PCr), inorganic phosphates (P_i) and sugar phosphates. (ADP is only present in small quantities and is anyway masked by an overlap with ATP.) Each class of healthy tissue contains characteristic proportions of these phosphates, giving rise to characteristic spectra. Cutting the blood flow to any tissue (ischaemia) alters the concentrations and rates of interconversion of these metabolites increasing the level of P_i while decreasing that of PCr, and there is a build-up of lactate, the byproduct of the regeneration of ATP *via* glycolysis, which decreases the pH.

The TMR technique has made great progress over the past year, encouraged by the success of the initial trials with test phantoms of ATP, PCr and P_i in solutions of different pH contained one within the other, which were followed by *in vivo* measurements on an anaesthetised rat in which the difference between its liver and the surrounding muscle was clearly visible. A fully developed system of horizontal superconducting magnet (operating in persistent mode and giving a field of 1.89 T), profiling coils and operating console is currently in operation in Oxford Instruments' application laboratory, and a second system is in the care of George Radda's group in the biochemistry department of Oxford University, a group that has played an important part in the development of TMR. Efficient (and speedy) data collection and processing are areas which are always in line for possible improvement — after all TMR can only give useful 'real time' information if it can operate on a shorter timescale than the metabolic changes it seeks to follow. However, the major challenge is now to medical research — to interpret the data that TMR supplies in physiological terms and to identify appropriate remedial action.

Appendix III   The Antecedents of Newport Instruments Limited

In the 1960s Mr Martin Wood, chairman of the Oxford Instruments Group, and Mr Ian Boswell, the original founder of Newport Instruments Limited, were known to each other through their connections with the Clarendon Laboratory, Oxford. During the same period, Mr Barrie Marson and Mr John Lee were working together at Kent Instruments, Luton. In 1970, Barrie Marson became managing director of Oxford Instruments Limited and, a year later, established a relationship with the Praxis Company of Italy and set up a UK subsidiary, Praxis Oxford, to market the Praxis Digital Systems in the UK. Praxis already had a subsidiary operation in Holland.

In 1974 Newport Instruments needed to strengthen their technical expertise with financial and commercial backing and for an exchange of shares Newport Instruments became a wholly owned subsidiary of the Oxford Instruments Group. At the same time, the Praxis Italian operation got into financial difficulties and was eventually taken over by the large Italian instrument group, Carlo Gavazzi. During a period when the Praxis factory in Italy was actually closed, order-book commitments were met by a co-operation between Oxford Instruments and Praxis Holland, the former providing manufacturing facilities and the latter carrying out engineering work.

During 1974 the electromagnet products of Newport Instruments were transferred to Oxford, leaving the other two activities of NMR Analysis Instruments and Wound Magnetic Components to expand at Newport Pagnell. Following the takeover of Praxis by Carlo Gavazzi, a UK subsidiary was formed, Carlo Gavazzi (UK) Limited, to replace the activities of Praxis Oxford Limited in marketing the digital systems and this was based in the Newport Instruments premises at Newport Pagnell. Oxford Instruments Limited took a minor shareholding in the Gavazzi subsidiary, and Barrie Marson invited John Lee to become its managing director.

In 1976 it was decided to manufacture the Gavazzi systems in the UK and, as the Newport Instruments factory had spare capacity, this unit was established at Newport Pagnell. A small team of systems engineers and specialists were recruited from the computer and process control field with the intention that this team should form the nucleus of a new joint manufacturing company between Oxford and Gavazzi. However, due to the economic climate in Italy at that time, the investment from Gavazzi was not forthcoming and, in 1977, this activity became a division of Newport Instruments Limited and Mr Lee became the managing director of the Newport Instruments company.

For two years Newport Instruments continued to manufacture and market the DIAC system (Distributed Intelligent Acquisition and Control System) under an agreement with the Gavazzi company, which was never wholly satisfactory, as a condition of the license was that Newport Instruments was excluded from the petrochemical market. A completely new system was therefore developed by the Newport systems team, known as System 86, and this was launched at the beginning of 1980 with considerable subsequent success.

Appendix IV Organization of Oxford Instruments Limited (180 employees)

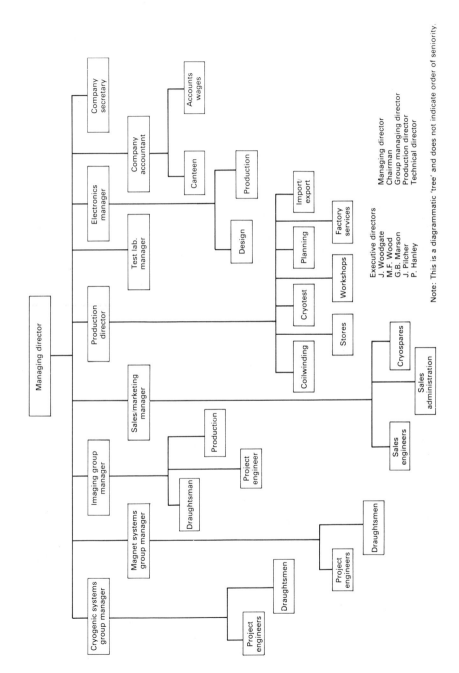

Managing director

Company secretary

Electronics manager

Company accountant

Canteen

Accounts wages

Production director

Test lab. manager

Production

Design

Import/export

Planning

Factory services

Workshops

Cryotest

Stores

Coilwinding

Sales/marketing manager

Imaging group manager

Draughtsman

Production

Project engineer

Magnet systems group manager

Cryogenic systems group manager

Project engineers

Draughtsmen

Project engineers

Draughtsmen

Sales engineers

Cryospares

Sales administration

Executive directors
J. Woodgate
M.F. Wood
G.B. Marson
J. Pilcher
P. Hanley

Managing director
Chairman
Group managing director
Production director
Technical director

Note: This is a diagrammatic 'tree' and does not indicate order of seniority.

Appendix V Organization of Oxford Medical Systems Limited (140 employees)

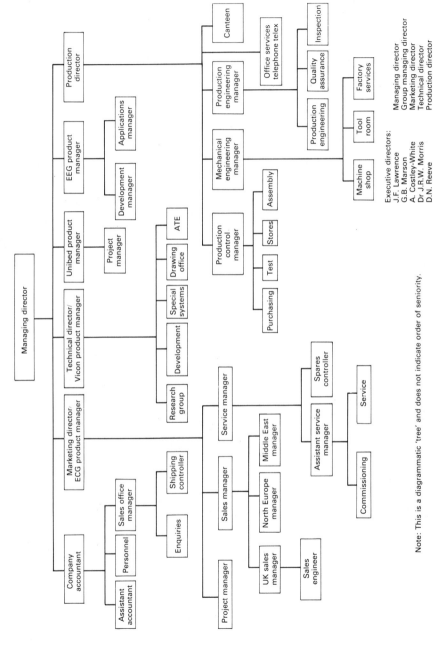

Managing director

- **Marketing director/ECG product manager**
  - **Company accountant**
    - Assistant accountant
    - Personnel
    - Sales office manager
      - Enquiries
      - Shipping controller
  - **Service manager**
    - **Project manager**
    - **Sales manager**
      - UK sales manager
        - Sales engineer
      - North Europe manager
      - Middle East manager
    - **Assistant service manager**
      - Commissioning
      - Service
    - **Spares controller**

- **Technical director/Vicon product manager**
  - **Research group**
    - **Development**
      - Special systems
      - Drawing office
      - ATE
  - **Unibed product manager**
    - **Project manager**
  - **EEG product manager**
    - **Development manager**
    - **Applications manager**
  - **Production control manager**
    - Purchasing
    - Test
    - Stores
    - Assembly

- **Production director**
  - **Production engineering manager**
    - Office services telephone telex
    - **Mechanical engineering manager**
      - Machine shop
      - Tool room
      - Production engineering
    - **Quality assurance**
      - Factory services
      - Inspection
  - **Canteen**

Executive directors:
J.F. Lawrence
G.B. Marson
A. Costley-White
Dr J.R.W. Morris
D.N. Reeve

Managing director
Group managing director
Marketing director
Technical director
Production director

Note: This is a diagrammatic 'tree' and does not indicate order of seniority.

Appendix VI Organization of Newport Instruments Limited (130 employees)

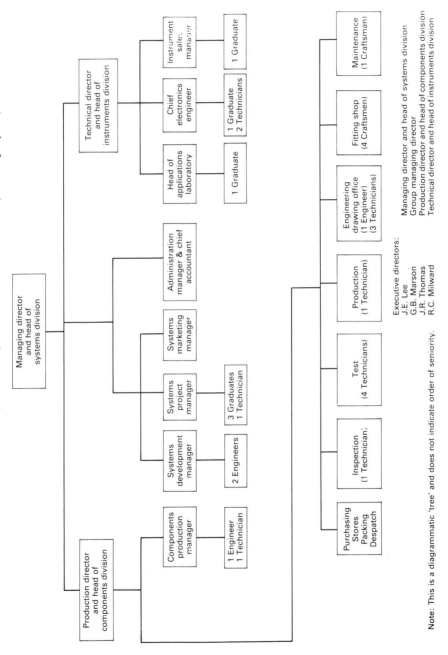

Note: This is a diagrammatic 'tree' and does not indicate order of seniority.

Executive directors:
J.E. Lee — Managing director and head of systems division
G.B. Marson — Group managing director
J.R. Thomas — Production director and head of components division
R.C. Milward — Technical director and head of instruments division

Appendix VII   Biographies of the Board Members

M.F. Wood   Age 54

Graduate in mechanical engineering from Cambridge University; subsequently worked in the Clarendon Laboratory Oxford
— built some of the world's first superconducting magnets
— demand for these magnets from all over the world led him to leave the university in 1962 and start the company.

G.B. Marson   Age 49

Graduate in physics from Nottingham University
— did research at College of Aeronautics Cranfield for two years and two years on flight-test instrumentation at De Havillands
— 15 years at Kent Instruments with the first ten in development and the last five as director of the digital systems division
— then came to Oxford Instruments.

J.M. Woodgate   Age 40

Engineering apprentice at Harwell; joined the company in 1965
— did development work on dilution refrigerators
— then became manager of the standard products section of the company as it was, then
— became managing director of Oxford Instruments Limited in 1976.

J.F. Lawrence   Age 41

— qualified as Chartered Accountant in the City
— subsequently had accounting jobs with IBM, Lotus Cars, Smiths of Witney
— joined the company as group financial director in 1972
— became managing director of Newport Instruments Limited for two years in 1975
— after the company was acquired became managing director of Oxford Medical Systems Limited in 1978.

J.E. Lee   Age 49

— Chartered mechanical engineer
— four years as engineer officer in the Navy
— subsequently worked for Kent Instruments for about 22 years principally on export sales
— ran the Kent Digital Systems office in Germany for three years
— Managing director of Carlo Gavazzi (UK) Limited from 1975 to 1978
— Managing director of Newport Instruments since 1978.

Appendix VIII   Product Range for Oxford Instruments Limited

Cooling Units

Product	Equipment factors	Temperature range	Applications
CF100 series	Supplied with gas flow pump and digital temperature controller. Range of transfer lines gives choice of cold-head orientation	3.5 K to 500 K	Optical, IR, and UV spectroscopy (CF104). X-ray diffraction (CF108)
CF200 series	Mounted in exchange gas and can be changed in seconds. Supplied with cryostat, transfer line, gas flow pump and controller, and digital temperature controller	3.6 K to 500 K	Experiments in either resistive or superconducting magnets. Hall effect, Mossbauer, and magnetic susceptibility experiments
CF500 series	Mounting as for CF200. Built-in helium transfer line and mounts directly on to helium storage vessel. Helium consumption much reduced so can be run for long periods at low cost		Mossbauer, neutron diffraction, and long-term X-ray experiments
ESR900	Supplied with spare tube transfer line, gas flow pump, and digital temperature controller	77–300 K (nitrogen) Down to 4 K (liquid helium)	Continuous flow system for ESR experiments. Compatible with most X-band cavities
ESR10	As for ESR900 but with additional pumped helium reservoir	Down to 1.8 K	Ditto
ESR35		4–400 K	Q-band work
MD series (MD4, MD8 and SMD8)	MD4 and MD8 available with variety of attachments, variable temperature inserts, etc. SMD range of standard open-top Dewars is available with helium reservoirs up to 200 mm diameter		Optical and X-ray experiments. Infra-red detector cooling
DN704	Low-cost variable temperature nitrogen cryostat		Optical and other experiments
K Jet	Produces stream of cold dry air. Can be used for experiments of long duration	Ambiant to −50°C	X-ray diffractometers and Weisenberg cameras

Magnets

Product	Equipment features	Technical data	Applications
NMR superconducting magnets	Solenoid is key to resolution. Low boil-off rate Dewar control critical item in running costs. Needs filling only four times a year. Autotracking for whole body NMR magnets	Full range 100–500 MHz	Studies of living animal and human systems
Spectromag	Compact enough to fit most commercial and purpose built spectrometers and spectropolarimeters. Spectromag 1 is fixed temperature instrument. Spectromags 2, 3, and 4 offer variable temperatures and a digital temperature controller is included in the standard system	Magnetic fields up to 6 tesla (8 tesla with large tails). Variable temperatures down to 1.5 K	Magneto-optical studies. Mossbauer spectroscopy. Faraday effect. Spin Flat Raman effect (SFR)
Electromagnets	Power supplies available providing wide range of fields and stabilities. Ancillary equipment includes rotary basis, high and low loading trolleys, pole tips, field modulating coils, and sweep generators	Magnetic fields up to 2.8 tesla Pole-faces up to 178 mm	
Magnetic susceptibility systems	Magnet with gradient coils, environmental control equipment, micro-balance, and data handling. Full pumping and control systems provided	Temperature 3.5–1000 K for electro-magnetic systems. Temperature 1.5–500 K for cryogenic magnets	Small magnets for Guoy experiments to high field (10T) superconducting magnets. Comprehensive systems including Faraday and Guoy facilities and full microprocessor data logging and environmental control

Low Temperature Equipment

Product	Equipment features	Temperature range	Applications
Dilution refrigerator	Three standard systems, identified by the cooling power (in microwatts at 100 mK) of 15 100 and 300 are offered. Larger units made up to 1000 mW can be specially built	Below 0.3 K	Low-temperature physics research

Appendix IX Product Range for Oxford Medical Systems Limited

Product	Technical features	Applications
Medilog 1	Four-channel 24-h cassette recorder of modular construction. Has plug-in amplifiers for different physiological signal conditioning	ECG. Work medicine. Sports medicine. EEG monitoring for absence. Seizures and sleep studies. Long-term apnoea monitoring and research
Medilog 2	Two-channel ECG cassette recorder incorporating all experience of Medilog 1. FM recording with flutter compensation provides clean recordings with flat base line and 0.05 LF response. Unique binary codes allow autosearch facility on replay. Autostop 'on event' facility on analyser for easy patient identification. Smallest and lightest recorder of its type results in high degree of patient acceptance. (Can be worn by small children, e.g. in hair, without detection)	
EEG analysers	One or two channels. Large display. Digital interface and microprocessor for variable format report writing. Wide beat detection and counting. Histograms display, count and write out	
Page mode display PMD 12	Replay speed up to sixty times recorded speed, so 24-h recordings replayed in 24 min	Replay and supply of up to four channels of EEG recorder on Medilog 1

Appendix X Product Range for Newport Instruments Limited

Product	Technical features	Applications
Ferrite cord transformers (766 series)	Lead-out wires of tinned copper or tinned phosphor bronze 0.02 mm diameter. Alternative values of turns ratio (1:1 to 4:1), ET constant (4 to 40, and inductance can be supplied	Suitable for working voltages up to 250 V d.c.
Pulse transformers (772 series)	Transformers impregnated and then encapsulated in high-grade epoxy case. Alternative values of turns ratio (1:1 to 2.1:1), ET constant (1.6 to 23), and inductance can be supplied	High-voltage insulation between windings
SCR and Triac triggering transformers (784 series)	Moulded case filled with high-grade epoxy resin. Alternative values of turns ratio ET constant, and inductance	To meet requirements for circular format with pins on a 0.1-in. grid

Product	Technical features	Applications
Inductor modules (803 series)	Fully encapsulated for maximum protection. Printed circuit board mounting. Low loss with excellent d.c. stability	Switching voltage regulators
MkIIIA and MkIIIF analysers	Rapid reading of hydrogen content of up to 150 ml. Magnet is 190 mm × 280 mm × 390 mm weighing 44 kg. Electronic console is 530 mm × 310 mm × 450 mm weighing 20 kg	Measuring hydrogen content in aviation fuel and residual oil in wax. Measuring plasticizer content in PVC and mobility in polyolefins. Measuring oil moisture in oil seeds, meat pasties, and chocolate
NMR process analyser (OA1)	Magnet/transducer unit contains magnet and RF coil assembly through which process material is piped and NMR absorption strength measured. Temperature sensors for monitoring temperature of flow sample. Control unit designed for remote control but full range of manual control for setting up. Automatic tuning, gain calibration, and compensation for long-term drift of NMR resonance signal. Sixteen individually addressable digi-switches by pre-programme microprocessor arithmetic. Digital and alphanumeric panel meters to display results	Remote on-line monitoring of hydrogen content of process materials. Analysis and quality control in food, polymer, and petrochemical industries. Measurement of fat content, solid/liquid ratio in edible fats, and moisture in foods. Measurement of plasticizer or rubber content of polymers and rigidity. Monitoring of distillation column and catalytic cracking column products. Measurement of moisture in coal and oil in explosives
Protein analyser (P100)	Based on pulsed nuclear magnetic resonance spectrometer in combination with aqueous relaxation agent. Relaxation rates measured in seconds, correlated with protein content, and directly displayed. Ergonomic design makes protein analyser easy to use, i.e. minimum of controls. Provides hired copy of experimental results	Fast, simple, and accurate method for the determination of protein and carbohydrate content
System 86	Master station containing powerful microprocessor for executive control of whole system. Master station communicates with remote stations along cable up to 5 km long. Remote units contained in small watertight boxes. RU 16 caters for sixteen analogue or digital inputs or outputs. RU 48 caters for sixteen analogue or sixteen digital inputs and sixteen digital outputs. RU microprocessor scans input/output modules and transfers data to RU memory for transference to MS when required	To provide industrial users with a reliable cost-effective way of improving plant reliability and productivity

Appendix XI Group Financial Performance and Staff Numbers, 1960–80

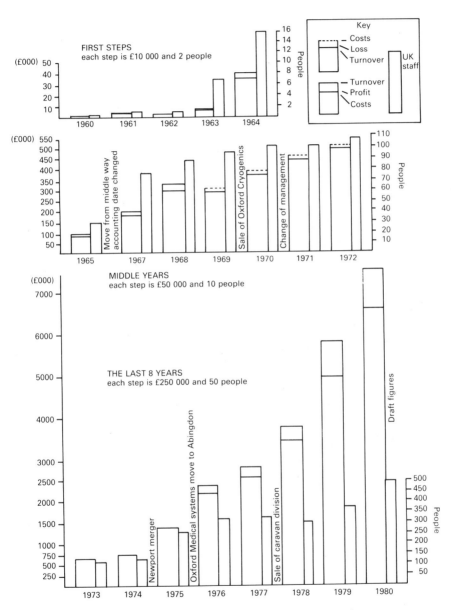

Some figures are adjusted to compare with present accounting practices

Appendix XII The Oxford Instruments Group

Consolidated Profit and Loss Account 52 Weeks Ended 25 March 1979

	1979 *£*	*1978* *£*
Turnover	5 756 627	3 733 607
Profit before taxation and extraordinary item	804 536	402 249
Dividend received from an associated company	8 832	–
	813 368	402 249
Taxation	464 036	204 390
Profit before extraordinary item	349 332	197 859
Extraordinary item	8 916	131 235
	358 248	329 094
Dividend	2 520	2 520
	355 728	326 574
Unappropriated profit brought forward	532 003	205 429
Unappropriated profit carried forward	887 731	532 003

Appendix XIII   The Oxford Instruments Group

Consolidated Balance Sheet 25 March 1979

	1979 £	*1978* £
Fixed assets	1 038 529	439 369
Subsidiary companies	51 653	7 394
Associated companies	1 769	–
	1 091 951	446 763
Current assets		
Stock and work in progress	1 962 357	1 034 667
Debtors	1 158 717	987 726
Amounts due from unconsolidated subsidiaries	316 510	104 623
Bank and cash balances	41 990	123 820
	3 479 574	2 250 836
Current liabilities		
Creditors	2 087 366	1 228 128
Bank overdraft (secured)	243 040	117 354
Taxation	108 652	124 354
	2 439 058	1 469 836
Net current assets	1 040 516	781 000
	2 132 467	1 227 763
Share capital	87 410	87 410
Share premium	109 883	109 883
Capital reserve	71 457	72 847
Unappropriated profit	887 731	532 003
	1 156 481	802 143
Deferred taxation	783 986	380 370
Secured debentures	192 000	45 250
	2 132 467	1 227 763

CHAPTER 11

Summary and Conclusions

11.1 Guidelines for Product Innovation

The design, manufacture, and marketing of new products are activities which involve all levels of management in every division of a company over a period of years. The most important part of product innovation lies in finding an idea which creates, or meets, a new market-need, and the more successfully this is achieved the easier will be all subsequent parts of the innovation process. The factor of second importance will be the direction of the development stages, for many will contain speculative elements and all will require completing within a rigorous timetable.

The diversity of new-product undertakings among companies is great, and it may be thought impossible to postulate management factors which are likely to better facilitate or hinder success. This was found not to be the case, and from a survey of relevant literature and from observations within eight companies three sets of guidelines on corporate behaviour and three sets on research, design, and development were compiled.

11.2 The Ways in Which Companies Innovate

All eight companies selected for this study were engaged upon and expressed enthusiasm for new-product development. Most of the companies practised some degree of corporate planning, and boards paid attention to the role of new products in their plans for growth. This was reflected in statements on company purpose, in lists of company strengths and weaknesses, and in the nature of both long- and short-term strategies. Research and development staff showed, by the way in which they organized their departments, that they, too, appreciated that evolutionary development needed to be supplemented by innovation.

Neither executive nor research staff, however, prepared long-term plans which were adequate in all respects. Little or no thought was given to the important decision of the allocation of funds for the future needs of research and development, marketing, or plant renewals. It was also difficult to find examples of projects being subject to quantitative estimates of development times, production costs, and eventual benefits from sales.

Companies' staff were generally unfamiliar with those current studies which are being made into the creative process throughout Europe and the USA, and consequently no instance was noticed of attempts being made to stimulate radical ideas deliberately. This casual attitude towards the quality of ideas led to research, design, and development staff neglecting to make use of helpful procedures and methodologies which were available. Furthermore, action was rarely taken to improve the ability of staff to recognize opportunities or to solve problems, nor was there an awareness of the crucial contribution to innovation which may result from the encouragement of informal information channels.

11.3 Model for Product Innovation

Nyström has examined in great detail the ways in which an innovative company differs from a positional one, and this study shows how his analysis may be extended. A distinction is drawn between six types of companies based upon the level of technology associated with the design and development of their products. Each level is defined by three parameters: the nature of the problem-solving activity, the qualifications and skills of the scientific and technical staff, and the merit of the research and development work judged by its suitability for publication. It is argued that the six levels of technology are associated with characteristics which determine the evolutionary and innovative behaviour of companies.

With the help of the model the formulation of innovative strategies becomes easier. For example, from a knowledge of the factors linked with a given level of technology it is possible to specify changes that will need to be made in human and material resources should a proposed merger, acquisition, or licensing arrangement involve a different technological level. Again, the model indicates ways in which an organization could be changed should it prove necessary to meet competition by raising the level of technology.

11.4 Conclusions

Successful companies which are engaged upon the manufacture of highly technical products are characterized by their employment of highly qualified staff who possess special aptitudes for solving problems, by their aggressive research and development tactics, by an organic organizational structure, and by reliance upon the production of relatively small numbers of high-value products. At the lowest scale of technology firms will employ few, if any, qualified scientists or technologists. There will be a formal, bureaucratic organization and emphasis will be placed upon corporate planning, while evolutionary development will be aimed at maintaining the economic manufacture of large production batches. Most companies, however, are concerned with the two middle levels of technology, and their situation is the most demanding since they must, somehow, combine features normally associated with the innovative and positional modes.

The case histories demonstrate that survival of a company can be threatened through neglect of a single management activity. At one period during its early history company No. 8 nearly failed through lack of commercial skills, and company No. 7 was recently disturbed by an internal conflict, the resolution of which was made difficult by the absence of a stable structure at board level.

A corollary of these observations is that successful innovation will only be accomplished when every management activity is carried out in a manner appropriate to the level of technology at which the company operates. It follows, therefore, that staff, and particularly scientists and technologists, must acquire and apply a wide range of skills, and for this they may need to develop special personal qualities. A lack of breadth has frequently been observed during this study (see, for example, page 177) and there is undoubtedly considerable potential improvement to be gained from wider application of existing knowledge. Action should be taken to design teaching or training modules suitable to each level of technology, whereby all subjects relevant to product development will be covered.

Scientists and technologists who work in manufacturing industries are faced with numerous challenges, most of which will be very demanding. Industry presents unique opportunities for contact not only with peers inside and outside industry but with production, sales, marketing staff, and customers. This will lead to an awareness of the varied circumstances which surround a problem and to a realization of the difficulty of discriminating between the relevant and irrelevant. It will, moreover, often be necessary to take decisions based upon only a partial understanding of the factors involved and to solve problems within given time-limits. The main stimulus is likely to be derived from competition, and the greatest sense of achievement from designing and manufacturing products which surpass all others. There will, however, be occasions when the success of competitors may seem to render months of strenuous work unavailing. This varied and exciting scenario will call for an element of aggression combined with patience, an ability to listen to and impart information, enthusiasm, and, above all, a toughness of character.

Index

n or befc